Reviews of Environmental Contamination and Toxicology

VOLUME 249

More information about this series at http://www.springer.com/series/398

Reviews of Environmental Contamination and Toxicology

Editor
Pim de Voogt

Founding Editor
Francis A. Gunther

Volume 249

Springer

ISSN 0179-5953 ISSN 2197-6554 (electronic)
Reviews of Environmental Contamination and Toxicology
ISBN 978-3-030-20196-8 ISBN 978-3-030-20194-4 (eBook)
https://doi.org/10.1007/978-3-030-20194-4

This Springer imprint is published by the registered company Springer Nature Switzerland AG
The registered company address is: Gewerbestrasse 11, 6330 Cham, Switzerland

Foreword

International concern in scientific, industrial, and governmental communities over traces of xenobiotics in foods and in both abiotic and biotic environments has justified the present triumvirate of specialized publications in this field: comprehensive reviews, rapidly published research papers and progress reports, and archival documentations These three international publications are integrated and scheduled to provide the coherency essential for nonduplicative and current progress in a field as dynamic and complex as environmental contamination and toxicology. This series is reserved exclusively for the diversified literature on "toxic" chemicals in our food, our feeds, our homes, recreational and working surroundings, our domestic animals, our wildlife, and ourselves. Tremendous efforts worldwide have been mobilized to evaluate the nature, presence, magnitude, fate, and toxicology of the chemicals loosed upon the Earth. Among the sequelae of this broad new emphasis is an undeniable need for an articulated set of authoritative publications, where one can find the latest important world literature produced by these emerging areas of science together with documentation of pertinent ancillary legislation.

Research directors and legislative or administrative advisers do not have the time to scan the escalating number of technical publications that may contain articles important to current responsibility. Rather, these individuals need the background provided by detailed reviews and the assurance that the latest information is made available to them, all with minimal literature searching. Similarly, the scientist assigned or attracted to a new problem is required to glean all literature pertinent to the task, to publish new developments or important new experimental details quickly, to inform others of findings that might alter their own efforts, and eventually to publish all his/her supporting data and conclusions for archival purposes.

In the fields of environmental contamination and toxicology, the sum of these concerns and responsibilitics is decisively addressed by the uniform, encompassing, and timely publication format of the Springer triumvirate:

Reviews of Environmental Contamination and Toxicology [Vol. 1 through 97 (1962–1986) as Residue Reviews] for detailed review articles concerned with any aspects of chemical contaminants, including pesticides, in the total environment with toxicological considerations and consequences.

Bulletin of Environmental Contamination and Toxicology (Vol. 1 in 1966) for rapid publication of short reports of significant advances and discoveries in the fields of air, soil, water, and food contamination and pollution as well as methodology and other disciplines concerned with the introduction, presence, and effects of toxicants in the total environment.

Archives of Environmental Contamination and Toxicology (Vol. 1 in 1973) for important complete articles emphasizing and describing original experimental or theoretical research work pertaining to the scientific aspects of chemical contaminants in the environment.

The individual editors of these three publications comprise the joint Coordinating Board of Editors with referral within the board of manuscripts submitted to one publication but deemed by major emphasis or length more suitable for one of the others.

Coordinating Board of Editors

Preface

The role of *Reviews* is to publish detailed scientific review articles on all aspects of environmental contamination and associated (eco)toxicological consequences. Such articles facilitate the often complex task of accessing and interpreting cogent scientific data within the confines of one or more closely related research fields.

In the 50+ years since *Reviews of Environmental Contamination and Toxicology* (formerly *Residue Reviews)* was first published, the number, scope, and complexity of environmental pollution incidents have grown unabated. During this entire period, the emphasis has been on publishing articles that address the presence and toxicity of environmental contaminants. New research is published each year on a myriad of environmental pollution issues facing people worldwide. This fact, and the routine discovery and reporting of emerging contaminants and new environmental contamination cases, creates an increasingly important function for *Reviews*. The staggering volume of scientific literature demands remedy by which data can be synthesized and made available to readers in an abridged form. *Reviews* addresses this need and provides detailed reviews worldwide to key scientists and science or policy administrators, whether employed by government, universities, nongovernmental organizations, or the private sector.

There is a panoply of environmental issues and concerns on which many scientists have focused their research in past years. The scope of this list is quite broad, encompassing environmental events globally that affect marine and terrestrial ecosystems; biotic and abiotic environments; impacts on plants, humans, and wildlife; and pollutants, both chemical and radioactive; as well as the ravages of environmental disease in virtually all environmental media (soil, water, air). New or enhanced safety and environmental concerns have emerged in the last decade to be added to incidents covered by the media, studied by scientists, and addressed by governmental and private institutions. Among these are events so striking that they are creating a paradigm shift. Two in particular are at the center of ever increasing media as well as scientific attention: bioterrorism and global warming. Unfortunately, these very worrisome issues are now superimposed on the already extensive list of ongoing environmental challenges.

The ultimate role of publishing scientific environmental research is to enhance understanding of the environment in ways that allow the public to be better informed or, in other words, to enable the public to have access to sufficient information. Because the public gets most of its information on science and technology from internet, TV news, and reports, the role for scientists as interpreters and brokers of scientific information to the public will grow rather than diminish. Environmentalism is an important global political force, resulting in the emergence of multinational consortia to control pollution and the evolution of the environmental ethic. Will the new politics of the twenty-first century involve a consortium of technologists and environmentalists, or a progressive confrontation? These matters are of genuine concern to governmental agencies and legislative bodies around the world.

For those who make the decisions about how our planet is managed, there is an ongoing need for continual surveillance and intelligent controls to avoid endangering the environment, public health, and wildlife. Ensuring safety-in-use of the many chemicals involved in our highly industrialized culture is a dynamic challenge, because the old, established materials are continually being displaced by newly developed molecules more acceptable to federal and state regulatory agencies, public health officials, and environmentalists. New legislation that will deal in an appropriate manner with this challenge is currently in the making or has been implemented recently, such as the REACH legislation in Europe. These regulations demand scientifically sound and documented dossiers on new chemicals.

Reviews publishes synoptic articles designed to treat the presence, fate, and, if possible, the safety of xenobiotics in any segment of the environment. These reviews can be either general or specific, but properly lie in the domains of analytical chemistry and its methodology, biochemistry, human and animal medicine, legislation, pharmacology, physiology, (eco)toxicology, and regulation. Certain affairs in food technology concerned specifically with pesticide and other food-additive problems may also be appropriate.

Because manuscripts are published in the order in which they are received in final form, it may seem that some important aspects have been neglected at times. However, these apparent omissions are recognized, and pertinent manuscripts are likely in preparation or planned. The field is so very large and the interests in it are so varied that the editor and the editorial board earnestly solicit authors and suggestions of underrepresented topics to make this international book series yet more useful and worthwhile.

Justification for the preparation of any review for this book series is that it deals with some aspect of the many real problems arising from the presence of anthropogenic chemicals in our surroundings. Thus, manuscripts may encompass case studies from any country. Additionally, chemical contamination in any manner of air, water, soil, or plant or animal life is within these objectives and their scope.

Manuscripts are often contributed by invitation. However, nominations for new topics or topics in areas that are rapidly advancing are welcome. Preliminary communication with the Editor-in-Chief is recommended before volunteered review manuscripts are submitted. *Reviews* is registered in WebofScience™.

Inclusion in the Science Citation Index serves to encourage scientists in academia to contribute to the series. The impact factor in recent years has increased from 2.5 in 2009 to 7.0 in 2017. The Editor-in-Chief and the Editorial Board strive for a further increase of the journal impact factor by actively inviting authors to submit manuscripts.

Amsterdam, The Netherlands Pim de Voogt
August 2018

Contents

Contributors

Saroj Arora Department of Botanical and Environmental Sciences, Guru Nanak Dev University, Amritsar, India

Shagun Bali Department of Botanical and Environmental Sciences, Guru Nanak Dev University, Amritsar, India

Ram Naresh Bharagava Laboratory for Bioremediation and Metagenomics Research (LBMR), Department of Environmental Microbiology (DEM), Babasaheb Bhimrao Ambedkar University (A Central University), Lucknow, Uttar Pradesh, India

Renu Bhardwaj Department of Botanical and Environmental Sciences, Guru Nanak Dev University, Amritsar, India

Herman Eijsackers Unit for Environmental Sciences and Management, North-West University, Potchefstroom, South Africa

Wageningen University & Research, Wageningen, The Netherlands

Robert M. Gogal Jr Department of Biosciences and Diagnostic Imaging, College of Veterinary Medicine, University of Georgia, Athens, GA, USA

Neha Handa Department of Botany, School of Bioengineering and Biosciences, Lovely Professional University, Phagwara, Punjab, India

David L. Haskins Savannah River Ecology Laboratory, University of Georgia, Aiken, SC, USA

Interdisciplinary Toxicology Program, University of Georgia, Athens, GA, USA

Warnell School of Forestry and Natural Resources, University of Georgia, Athens, GA, USA

Kersey Kalubi Biomolecular Sciences Program, Laurentian University, Sudbury, ON, Canada

Kanika Khanna Department of Botanical and Environmental Sciences, Guru Nanak Dev University, Amritsar, India

Sukhmeen Kaur Kohli Department of Botanical and Environmental Sciences, Guru Nanak Dev University, Amritsar, India

Mark Maboeta Unit for Environmental Sciences and Management, North-West University, Potchefstroom, South Africa

Melanie Mehes-Smith Department of Biology, Laurentian University, Sudbury, ON, Canada

Sikandar I. Mulla CAS Key Laboratory of Urban Pollutant Conversion, Institute of Urban Environment, Chinese Academy of Sciences, Xiamen, People's Republic of China

Ramya Narendrula-Kotha Biomolecular Sciences Program, Laurentian University, Sudbury, ON, Canada

Kabwe Nkongolo Biomolecular Sciences Program, Laurentian University, Sudbury, ON, Canada

Department of Biology, Laurentian University, Sudbury, ON, Canada

Diane Purchase Department of Natural Sciences, Faculty of Science and Technology, Middlesex University, London, UK

Adriaan Reinecke Department of Botany and Zoology, Stellenbosch University, Matieland, South Africa

Sophié Reinecke Department of Botany and Zoology, Stellenbosch University, Matieland, South Africa

Ganesh Dattatraya Saratale Department of Food Science and Biotechnology, Dongguk University-Seoul, Goyang-si, Gyeonggi-do, Republic of Korea

Gaurav Saxena Laboratory for Bioremediation and Metagenomics Research (LBMR), Department of Environmental Microbiology (DEM), Babasaheb Bhimrao Ambedkar University (A Central University), Lucknow, Uttar Pradesh, India

Anket Sharma State Key Laboratory of Subtropical Silviculture, Zhejiang A&F University, Hangzhou, China

Gabriel Theriault Biomolecular Sciences Program, Laurentian University, Sudbury, ON, Canada

Tracey D. Tuberville Savannah River Ecology Laboratory, University of Georgia, Aiken, SC, USA

Metal Toxicity and Resistance in Plants and Microorganisms in Terrestrial Ecosystems

Ramya Narendrula-Kotha, Gabriel Theriault, Melanie Mehes-Smith, Kersey Kalubi, and Kabwe Nkongolo

Contents

R. Narendrula-Kotha · G. Theriault · K. Kalubi
Biomolecular Sciences Program, Laurentian University, Sudbury, ON, Canada
e-mail: RY_narendrula@laurentian.ca; GX_theriault@laurentian.ca; KN_kalubi@laurentian.ca

M. Mehes-Smith
Department of Biology, Laurentian University, Sudbury, ON, Canada
e-mail: MS_mehes@laurentian.ca

K. Nkongolo (✉)
Biomolecular Sciences Program, Laurentian University, Sudbury, ON, Canada

Department of Biology, Laurentian University, Sudbury, ON, Canada
e-mail: knkongolo@laurentian.ca

P. de Voogt (ed.), *Reviews of Environmental Contamination and Toxicology, Volume 249*, Reviews of Environmental Contamination and Toxicology Volume 249,
https://doi.org/10.1007/398_2018_22

Abbreviations

γ-GC	Gamma-glutamylcysteine
ABC	ATP-binding cassette
ACC	1-Aminocyclopropane-1-carboxylic acid deaminase
Ag	Silver
AM	Arbuscular mycorrhizal fungi
APX	Ascorbate peroxidase
As	Arsenic
ATP	Adenosine triphosphate
CAT	Catalases
Cd	Cadmium
CDF	Cation diffusion facilitator
CEC	Cation exchange capacity
Co	Cobalt
COPT	Copper transporter
Cu	Copper
CzcABC	Cation-proton antiporter
ECM	Ectomycorrhizal fungi
EPS	Exopolysaccharides
Fe	Iron
GCS	γ-Glutamyl-Cys synthetase
GR	Glutathione reductase
GS	Glutathione synthetase
GSH	Glutathione
GST	Glutathione-s-transferase
H_2O_2	Hydrogen peroxide
Hg	Mercury
$HgCl_2$	Mercuric chloride
HMA	Heavy metal ATPase
IREG	Iron-regulated proteins
IRT	Iron-regulated transporter
K	Potassium
Mg	Magnesium
Mn	Manganese
MRP	Multidrug resistance-associated proteins
MT	Metallothionein
MTP	Metal tolerance proteins
NA	Nicotianamine
Na	Sodium
NADPH	Nicotinamide adenine dinucleotide phosphate
NAS	Nicotianamine synthase
Ni	Nickel
NRAMP	Natural resistance-associated macrophage proteins

O_2	Oxygen
OAS	O-acetyl-l-serine
OH	Hydroxyl radical
Pb	Lead
PC	Phytochelatin
ROS	Reactive oxygen species
S	Sulfur
SAT	Serine acetyltransferase
SOD	Superoxide dismutase
TMP	Putative transporter protein
Trr	Thioredoxin reductase
Ur	Uranium
ZAT	Zinc finger protein
ZIP	ZRT, IRT-like proteins
Zn	Zinc

1 Introduction

Metal pollution of soil and water is one of the most important environmental problems of the industrialized countries, affecting human health, agriculture, and forest ecosystem (Hohl and Varma 2010; Sherameti and Varma 2015).

Some metals can disrupt the proper functioning of pigments or enzymes by replacing essential elements (Henry 2000; Kushwaha et al. 2016). Long term exposure to metals can lead to health issues such as mental lapse, kidney and liver malfunction, gastro-intestinal tract, and central nervous systems disorders in humans (USDA NRCS 2000; Kushwaha et al. 2016). In plants, increased accumulation of metals found in soil is toxic. Their roots can absorb metal ions present in the soil in elevated concentration where they can then be translocated to their aerial biomass affecting important metabolisms (Bingham et al. 1986; Foy et al. 1978; Kushwaha et al. 2016). Moreover, high concentrations of metals in soil lead to a loss of soil fertility and agricultural yield, as well as a decrease in soil microbial activity (McGrath et al. 1995; Sherameti and Varma 2015; Kushwaha et al. 2016).

Metal toxicity depends on the element itself and its total/bioavailable concentrations for different organisms (Sherameti and Varma 2015; Kamal et al. 2010; Ahmad et al. 2005). The more mobile the metal is, the greater the exposure associated with it (Bourg 1995; Mulligan et al. 2001). Often, only a small portion of total metal in soil is available for uptake by soil microorganisms and plants. This fraction is present in the soil as either (1) free ions, (2) soluble complexes, or (3) readily desorbable (labile) forms (Alloway 2014). Many factors affect the availability of metals in soils including pH, redox state, soil texture, clay content, organic matter, oxides, and the presence of other cations in the soil. These soil properties determine the soil's ability to retain and immobilize metals (Aydinalp and Marinova 2003). Further, plants can also affect the availability of metals by releasing exudates from their roots.

Metal toxicity in plants is not as studied as in other organisms such as microorganisms and animals. The most recent review on metal resistance mechanisms in plants was completed by Singh et al. (2015). They focused only on the role of transcriptomics, proteomics, metabolomics, and ionomics. Later, Chandrangsu et al. (2017) reported a comprehensive review of metal homeostasis and resistance in bacteria. They also described transcriptional response to metal stress and posttranscriptional regulatory systems. On the other hand, Abdu et al. (2017) reported the effects of heavy metals on soil microbial population, diversity, and activities. This review summarizes key findings on metal toxicity and response in plants and decrypts resistance mechanisms involved in bacteria, archaea, and fungi. Genes associated with metal resistance mechanisms are described in details.

2 Metal Toxicity and Resistance in Plants

2.1 *Effects of Metals on Plants*

A metal is toxic when its concentration exceeds a specific threshold in a plant and it is dose-dependent (Hohl and Varma 2010; Burton et al. 1983; Dixon and Buschena 1988). The accumulation of metals within plant tissues at toxic thresholds can have detrimental effects on plants including growth retardation, shorter and more compact root systems and smaller leaf area, the browning of roots, foliar chlorosis, and necrosis. The degree of severity depends on several factors which include the metal tolerance of the plant, the bioavailable concentration of the metal, and the duration of exposure. Further, the response to metal toxicity differs by metal (Cuypers et al. 2013). For example, high bioavailable nickel (Ni) in the soil can lead to reduced plant biomass and cause leaf deformities, necrosis, as well as chlorosis. The activity of certain enzymes, a reduction in the uptake and translocation of certain nutrients, and a decrease in cell division have also been seen in plants in the presence of elevated nickel levels in soil. Further, copper (Cu) also has detrimental effects on plants if present in high concentration in the soil. Toxicity symptoms include root tissue damage, reduced photosynthetic activity, and increased permeability of the root cell plasma membrane as well as damage to DNA (Mehes-Smith and Nkongolo 2015).

Presence of excess amounts of metals has been also shown to cause acute toxic effects resulting in DNA damage (Hall 2002; Kamal et al. 2010; Hughes and Poole 1989). Metals can bind to proteins, leading to inhibition of activity or disrupting protein structure, and/or metals can displace essential elements resulting in deficiency effects (Hall 2002). Elevated levels of metals can stimulate formation of free radicals and reactive oxygen species (ROS) such as singlet oxygen (O_2), hydrogen peroxide (H_2O_2), and hydroxyl radicals (–OH) that cause oxidative stress resulting in cellular damage in organisms (Hall 2002; Emamverdian et al. 2015).

2.2 Copying Mechanisms in Plants to Elevated Soil Metal Concentrations

Plants have different strategies to cope with metal and prevent phytotoxicity once they access cellular or intercellular compartments. Depending on the metal and organisms, different modes of actions have been identified including binding to macromolecules (DNA, RNA, and proteins), disruption of enzymatic activity, and radical formation (Sherameti and Varma 2015; Winterhalder 1996; Gunn et al. 2007; Kamal et al. 2010). Plants prevent phytotoxicity by various processes which include formation of insoluble crystals and vesicles and retention of elements by cell walls (Mehrotra 2005). Further, metal tolerance is significantly influenced by accumulation of metals in root cells (Hall 2002; Hohl and Varma 2010). Plants have mechanisms to prevent oxidative stress, as well as enzymes and other compounds, to avoid cell damage by inhibiting or quenching free radicals and ROS.

2.2.1 Physiological Mechanisms

Resistant plants growing on metal-contaminated soils have developed physiological mechanisms to prevent the entry of metals in root cells as well as to cope with the effects of high internal metal concentrations. These coping strategies allow establishment of plant communities in metal-contaminated soils.

Plants are resistant to metals by either avoidance and/or tolerance strategies.

Avoidance: metal avoider plants use this strategy that prevents the entry of metal ions in their cells. They are able to extend their roots in the soil to explore less contaminated areas (Mehes-Smith et al. 2013). They can also prevent the uptake of bioavailable metals in their roots by immobilizing these ions by secreting root exudate in the rhizosphere. In several polluted soils, mycorrhizal fungi can prevent metal stress in their host plants by reducing the accumulation of metals in plant roots (Mehes-Smith et al. 2013). They can do this by extending their hyphae outside the plant's rooting zone and transferring the required elements to plants. Mycorrhizal fungi can also contribute to plant metal resistance by alternating metal speciation as well as preventing the entry of metals into plant roots. Mycorrhizal fungi can also chelate metal ions that have entered their cells with organic acids, phytochelatins, and polyphosphates and then transport them in their vacuoles (Cuypers et al. 2013; Mehes-Smith et al. 2013; Emamverdian et al. 2015). When the avoidance strategy is not achieved, tolerance mechanisms are activated.

Tolerance: Tolerant plant strategy consists in detoxifying metal ions that have entered their cell by crossing the plasma membrane or the biomembranes of internal organelles. They can be further classified into three categories. Excluders maintain a low level of metals in their aerial tissues when growing on metal-contaminated soils. Metals enter roots cells but they prevent their movement from roots to aerial tissues. Indicators accumulate metals in their aboveground tissues, but the levels found in these areal tissues are reflective of the metal concentration in the soil (the metal concentration in their aboveground tissue is similar to the levels found in the soil). In accumulators/hyperaccumulators, metals that have entered in roots are translocated and accumulated in their areal tissues. The levels of metals found in their harvestable tissues are higher than the levels found in the soil (Cuypers et al. 2013; Mehes-Smith et al. 2013).

In addition to these strategies, there are three main mechanisms used by plants to cope with metal contamination. They include restriction of metal uptake and transport, complexation of metals at the cell wall-plasma membrane interface, and complexation and compartmentation of metals within the cells (Das and Jayalekshmy 2015).

Restriction of Metal Uptake and Transport

Some plants restrict the entry of metals into their cells, while others don't. In general, plants restrict the entry of excess bioavailable metals by precipitating or by complexing them in the root environment. Methods of precipitation might include increasing the pH of the rhizosphere or the excretion of anions (e.g., phosphate) and organic acids (e.g., malate, citrate, oxalate) making them unavailable for uptake (Cuypers et al. 2013; Mehes-Smith et al. 2013).

Complexation of Metals at the Cell Wall-Plasma Membrane Interface

Metals can enter root tissues through either the symplast or through the apoplast. The first plant structure to come in contact with metals in soil is the plasma membrane of root cells. Its function is to act as a barrier to restrict the entry of metals into the cytoplasm (Das and Jayalekshmy 2015). High amounts of metal ions can also be found at the wall-plasma membrane interface. This can be achieved by altering the ion binding capacity of the cell wall. The cation exchange capacity (CEC) of metal-tolerant and metal-sensitive plant varies. Tolerant plant cell walls have higher CEC where metal ions can complex thus preventing their entry into the cell (Das and Jayalekshmy 2015).

Complexation and Compartmentation of Metals with the Cells

Studies have shown that plants have a range of mechanisms at cellular levels that are involved in metal uptake (Hall 2002; Emamverdian et al. 2015). These mechanisms consist in removing metals from the cytosol and/or sequestering the ions into a vacuole. Once inside the vacuole, the toxic effects of metals on sensitive organelles are reduced (Tong et al. 2004; Yadav 2010). The uptake of metals within the vacuole can be achieved by vacuolar pumps. Metal ions enter cells by competing with essential ions of similar ionic radii and with the help of proteins [phytochelatins

(PC) and metallothioneins (MT)] and transporters (Hall 2002; Emamverdian et al. 2015). Both PC and MT are cysteine-rich metal-chelating peptides capable of binding to various metals including cadmium (Cd), Cu, zinc (Zn), and arsenic (As) (Hall 2002; Saraswat and Rai 2011). Chelation of metals can restrict them within the cell or allow their sequestration and compartmentation. The production of PCs can be induced in the presence of metals such as Cu, Cd, Zn, Ag (silver), Hg (mercury), and Pb (lead) in roots, shoots, and leaves of tolerant plants such as in *Sedum alfredii*. Cd-PC complexes can accumulate in vacuoles, thus lowering the Cd ion concentration in the cytosol (Das and Jayalekshmy 2015; Mehes-Smith et al. 2013). Tolerant plants can also prevent the phytotoxicity effects of free metal ions within the cell by chelating them with specific ligands. Metal compartmentalization can also take place in the apoplast as well as specific cells of the epidermis and trichome (Yruela 2005).

2.3 Genes Involved in Metal Resistance in Plants

Metal-resistant genes code for proteins that play a direct role in the detoxification of either metals or the reactive oxygen species. Detailed description of genes associated with Ni and copper resistance in model and non-model plant species is described in Djeukam et al. (2016), Theriault et al. (2016a, b), and Makela et al. (2016). Theriault et al. (2016a) analyzed *Betula papyrifera* root transcriptome and reported that the main mechanism involved in nickel resistance is a downregulation of genes associated with translation (in ribosome), binding, and transporter activities. The majority of these genes can be classified into two distinct categories, chelators or metal transporters.

2.3.1 Chelators

Chelators are molecules that will sequester metals and render them inactive. They are generally small and low-carbon cost molecules that do not utilize much energy from plants when produced. Many chelators are byproducts from the Krebs cycle or other metabolic pathways. For example, increased cytosolic levels in nicotianamine (Mari et al. 2006), citrate (Ryan et al. 2009), malate (Liu et al. 2009), and amino acids such as proline (Sharma and Dietz 2009), histidine (Krämer et al. 1996), and glutathione (GSH) (Freeman et al. 2004) have been reported to play a role in metal resistance in plants. Many hyperaccumulators depend on chelating substrates for proper metal storage. This is the case for Ni hyperaccumulator *Sebertia acuminata*, where Ni was only accumulated in the aerial tissues when bound to citrate (Lee et al. 1977).

Genes involved in the production of these metabolites are often upregulated in metal-resistant plants. High levels of GSH, cysteine, and O-acetyl-l-serine (OAS) have been reported in the shoots of Ni hyperaccumulator *Thlaspi* spp. (Freeman et al. 2004). Increased levels of these metabolites have been shown to be correlated with high expression levels of both serine acetyltransferase (SAT) and glutathione reductase. These enzymes play an important role in the production and turnover of GSH (Freeman et al. 2004). High levels of manganese (Mn), iron (Fe), Cu, Ni, and Zn have also been correlated with increased expression of nicotianamine synthase (NAS) and nicotianamine production (Kim et al. 2005; Van de Mortel et al. 2006). NAS is an enzyme that plays a role in the synthesis of nicotianamine by the trimerization of S-adenosylmethionine (Shojima et al. 1990).

As previously discussed, larger molecules such as MTs and PCs have been extensively studied and known to play a role in metal detoxification and resistance (Hall 2002). MTs bind metals or reactive oxygen species through their thiol groups (Zhu et al. 2009). MTs play a role in general metal homeostasis; however, their involvement in metal resistance can't be overlooked (Hall 2002). They can be found in organisms from all kingdoms. MTs can be divided into three classes based on their cysteine arrangement. Class I MTs are found in vertebrates and have highly conserved cysteine arrangements, while Class II MTs are less conserved and is primarily found in plants, fungi, and invertebrates. Class II MTs can be further divided into four more categories, ranging in what tissues they are found in (Cobbett and Goldsbrough 2002). Class III MTs (phytochelatins) are exclusively found in plants (Prasad 2014). High expression of MTs has been correlated with resistance to Cu (Rauser and Curvetto 1980; Schultz and Hutchinson 1988) and Cd (Hasegawa et al. 1997; Lee et al. 2004) in plants.

Unlike MTs, genes do not directly encode PCs. Their structure consists of large oligomers of GSH. Like GSH, PCs can bind and immobilize metals through their thiol groups (Cobbett and Goldsbrough 2002). The PC pathway consists of three steps: (1) γ-glutamyl-Cys synthetase (GCS) catalyzes the dipeptide gamma-glutamylcysteine (γ-GC) from cysteine and glutamate, (2) glutathione synthetase (GS) produces GSH from γ-GC and an additional glutamate, and (3) phytochelatin synthase ligates multiple GSH molecules to form long oligomer chains (Cobbett and Goldsbrough 2002; Grill et al. 1989). The cad2-1 *Arabidopsis* mutant is deficient in GCS (Cobbett et al. 1998). Cad2-1 mutants have low levels of PCs and high sensitivity to Cd. This shows the importance of PCs in metal homeostasis (Cobbett et al. 1998).

2.3.2 Transporters

The second class of metal-resistant genes codes for large proteins called transporters. Transporters are involved in shuffling metals in different areas of the plant. In hyperaccumulators, excluders, and indicators, transporters are responsible for compartmentalizing metals in specific tissues. Transporters are often dependent on metal chaperones or other ions for proper functionality (Hall 2002). For instance, antiporters pump both ions across the membrane in opposite directions in contrast to symporters that pump both ions in the same direction. Some metal transporters that have been known to play a role in metal resistance include NRAMP (natural resistance-associated macrophage proteins), heavy metal ATPase (HMAs), cation diffusion facilitator (CDF), ABC transporter (ATP-binding cassette), and the ZIP (ZRT, IRT-like proteins) family. They are responsible for maintaining proper metal homeostasis in bacteria, animals, and plants (Viehweger 2014). Many of these transporters are non-specific and are able to transport a class of metals that have similar physical properties. Transporters from the IRT1 (iron-regulated transporter 1) family normally transport Fe but under deficient conditions can transport Zn, Ni, Cd, Co (cobalt), and Mg (magnesium) (Vert et al. 2002). Low transport specificity is also seen in *Arabidopsis thaliana*, where Ni and Zn compete for the AtZIP4/TcZNT1 transporter (Hassan and Aarts 2011). In *Betula papyrifera*, six candidate genes associated with nickel resistance were identified (Theriault et al. 2016a; Theriault and Nkongolo 2017). They include glutathione-s-transferase (GST), thioredoxin family protein, putative transmembrane protein, Nramp transporter, TonB receptor, and TonB dependent protein.

2.3.3 Specific Genes Associated with Nickel, Copper, and Cadmium Resistance in Plants

Genes that have been hypothesized to play a role in Ni resistance include glutathione reductase (GR), glutathione-s-transferase, serine acetyltransferase (SAT), nicotianamine synthase (NAS3), metal transporter (NRAMP), 1-aminocyclopropane-1-carboxylic acid deaminase (ACC), high affinity Ni transporter family protein (AT2G16800), Zn transporter of *Arabidopsis thaliana* (ZAT11), iron-regulated protein (IREG), thioredoxin family protein, and putative transmembrane protein (TMP) (Table 1).

Glutathione reductase (GR) is a metal chelator that contains a thiol and carboxylic group allowing its formation with Ni and other metals. Once the complexes are formed, they are sequestered into vacuoles by ABC-type transporters (Viehweger 2014). In Ni hyperaccumulator *Thlaspi goesingense*, increased levels of Ni in plants have been demonstrated to cause the overexpression of an enzyme called GR involved in the maintenance of high levels of GSH. This increase in GSH results in an increased resistance to the growth inhibitory- and oxidative stress-induced effects of Ni (Freeman et al. 2004; Theriault et al. 2016b; Makela et al. 2016; Kalubi et al. 2018).

Glutathione-s-transferases (GST) are a family of enzymes that catalyze the conjugation of electrophilic substrates to reduce GSH. The resulting complexes are transported to a vacuole for further processing or degradation. The removal of GSH facilitates the metabolism, sequestration, or removal of xenobiotic in roots. Recent studies have discovered increases in GST activity in root and shoot tissues in response to Ni toxicity in wheat (*Triticum aestivum*). More specifically, Ni was found to activate an isoform of glyoxalase I, which plays an important role in the degradation of methylglyoxal (MG). During stress, detoxification of MG occurs, and GSH levels are reduced and regenerated (Gajewska and Skłodowska 2008; Theriault et al. 2016b; Makela et al. 2016; Kalubi et al. 2018).

Table 1 Candidate genes involved in nickel resistance in model and non-model plant species

Gene	Species	Reference
1-Aminocyclopropane-1-carboxylic acid deaminase (ACC)	*Brassica napus*	Stearns et al. (2005)
High affinity nickel transporter family protein (AT2G16800)	*Arabidopsis thaliana*	Stearns et al. (2005)
The family of iron-regulated proteins (IREG)	*Arabidopsis thaliana*	Schaaf et al. (2006)
Glutathione reductase (GR)	*Thlaspi goesingense*	Freeman et al. (2004)
Glutathione-s-transferases (GST)	*Thlaspi goesingense* *Betula papyrifera*	Freeman et al. (2004) Theriault et al. (2016a)
Metal transporter NRAMP (1–4)	*Thlaspi japonicum* *Noccaea caerulescens* *Thlaspi caerulescens* *Betula papyrifera*	Mizuno et al. (2005) Visioli et al. (2012) Wang et al. (2008) Theriault et al. (2016b)
Nicotianamine synthase (NAS3)	*Noccaea caerulescens* *Thlaspi goesingense*	Visioli et al. (2012) Mari et al. (2006)
Putative transporter protein (TMP)	*Betula papyrifera*	Theriault et al. (2016a)
Serine acetyltransferase (SAT)	*Thlaspi goesingense*	Freeman et al. (2004)
Thioredoxin	*Chlamydomonas reinhardtii* *Betula papyrifera*	Lemaire et al. (2004) Theriault et al. (2016b)
TonB-dependent receptor/protein	*Betula papyrifera*	Theriault and Nkongolo (2017)
Zinc finger protein (ZAT11)	*Arabidopsis thaliana*	van der Zaal et al. (1999)

High activity of serine acetyltransferase (SAT) is involved in Ni resistance in the Ni hyperaccumulator *T. goesingense*. SAT is responsible for catalyzing the acetylation of L-serine to produce O-acetyl-L-serine (OAS), a positive regulator of sulfur (S) assimilation and an intermediate in the biosynthesis of the amino acid cysteine. The upregulation of SAT in *T. goesingense* has been associated with this species ability to hyperaccumulate Ni and to resist damaging oxidative effects (Freeman et al. 2004; Theriault et al. 2016b; Makela et al. 2016; Kalubi et al. 2018).

Nicotianamine synthase (NAS3) is an enzyme responsible for the synthesis of nicotianamine (NA) and plays a role in Ni tolerance in plants. Studies have revealed that exposure to Ni triggers accumulation of NA-Ni chelates in the xylem in the Ni hyperaccumulator *Thlaspi caerulescens* in a Ni dose-dependent manner. Once in the roots, sodium (Na) bound to Ni is redirected to aerial parts from root-to-shoot (Mari et al. 2006; Theriault et al. 2016b; Makela et al. 2016; Kalubi et al. 2018).

The family of natural resistance-associated macrophage proteins (NRAMP) is a group of metal ion transporters, which are an integral part of the cell membrane. NRAMP proteins help with the transportation of metals including Mn^{2+}, Zn^{2+}, Cu^{2+}, Fe^{2+}, Cd^{2+}, Ni^{2+}, and Co^{2+}. Homologues of this gene family have been found in several organisms including bacteria, insects, mammals, plants, and yeast (Williams et al. 2000). Mizuno et al. (2005) established the role of NRAMP transporters in Ni^{2+} transportation and homeostasis.

1-Aminocyclopropane-1-carboxylic acid deaminase (ACC) is an enzyme present in plant growth-promoting bacteria and is the biosynthetic precursor of the plant hormone ethylene. ACC has been demonstrated to lower plant ethylene levels and decrease the negative effects of various environmental stressors including Ni toxicity and contributing to increased Ni tolerance in canola (*Brassica napus*) (Stearns et al. 2005; Theriault et al. 2016b; Makela et al. 2016).

AT2G16800 is a known nickel/cobalt ion transporter from the NiCoT protein family. This protein utilizes the proton-motive force to drive the uptake of Ni^{2+} and Co^{2+} ions into the plant root system (Rodionov et al. 2006). This family of transporters is well characterized in prokaryotes that use these genes for nickel bioaccumulation (Deng et al. 2013). Though knowledge of the Ni/CoT is limited in plants, it is a good candidate gene involved in metal resistance in accumulators. High affinity transporter family protein (AT2G16800) binds Ni and is involved in transportation of Ni across the plasma membrane in the *A. thaliana* species (Stearns et al. 2005; Czajka et al. 2018).

Zinc finger protein (ZAT11) is a C_2H_2 zinc finger protein of *A. thaliana* and a nuclear-localized transcriptional regulator. ZAT11 positively regulates primary root growth at normal conditions and negatively regulates Ni^{2+} tolerance at excess levels of Ni. Mechanisms of reduced tolerance of Ni^{2+} is believed to involve the repression of transcription of a vacuolar Ni^{2+} transporter gene (Liu et al. 2014; Theriault et al. 2016b; Makela et al. 2016).

The family of iron-regulated proteins (IREG) is involved in the transport of transition metals from long distances in the *A. thaliana* species. IREG genes have long been known to encode Fe transporters; however, recent studies have discovered that IREG has low substrate specificity. Thus, under Fe deficiency, accumulation of other transition metals including Ni occurs where the metal is then transported into the vacuole (Schaaf et al. 2006).

Thioredoxin reductase (Trr1) plays a vital role in the reduction-oxidation system of thioredoxin. Following oxidative stress, this enzyme catalyzes the reduction of thioredoxin by NADPH (nicotinamide adenine dinucleotide phosphate). For this reason, Trr1 plays a critical role in protection and detoxification of cells experiencing oxidative stress as a result of metal toxicity. More specifically, Trr1 is important in cellular defense against Ni-induced DNA damage via a knockdown system of Trr1 using small interfering RNA (Kim and Seo 2012).

Putative transporter protein (TMP) has been established to be associated with Ni resistance during the transcription analysis of white birch (*Betula papyrifera*) (Theriault et al. 2016a, b). Its exact function is still unknown.

For Cu, a summary of important genes is described in Table 2. Some transporter families such as COPT (Cu transporter) and ZIP are involved in the transport of Cu to the intracellular space (Burkhead et al. 2009). Intracellularly, Cu can be

Table 2 Candidate genes involved in copper resistance in model and non-model plant species

Gene	Species	Reference
Response to antagonist 1 (RAN1)	*Arabidopsis thaliana*	Kobayashi et al. (2008)
Multidrug resistance-associated protein (MRP4)	*Betula pendula*	Keinanen et al. (2007)
Copper transporter protein (COPT 1)	*Arabidopsis thaliana*	Sancenon et al. (2004)
Metallothionein (MT2B)	*Arabidopsis thaliana* *Silene vulgaris* *Silene paradoxa*	Guo et al. (2008) van Hoof et al. (2001) Mengoni et al. (2003)

Response to antagonist 1 (RNA1): belongs to the P-type ATPase Cu-transporters
Multidrug resistance-associated protein (MRP4): subfamily of ATP-binding cassette (ABC) transporters, which are ATP-dependent and an integral part of plant detoxification
Metallothionein (MT2B): cysteine-rich proteins that bind metals including Cd, Cu, and Zn. MTs protect plants from oxidative stress by detoxifying hydroxyl radicals
Copper transporter protein (COPT1) belongs to a family of putative *Arabidopsis* copper transporters (COPT1-COPT5)

transmitted inside organelles by a variety of proteins. A particular family of transporter, the ATPases for heavy metals (HMA), is important at this level. Through ATP (adenosine triphosphate), these membrane transporters have a specific function of transporting metals through biological membranes (Baloun et al. 2014). At present, eight transporters of the HMA family (HMA1–HMA8) have been identified in *A. thaliana*. HMA1–HMA4 have been shown as playing a role in the transport of Zn, Cd, and Pb, while HMA5–HMA8 carry Hg and Cu. A specific transporter of this family, HMA7 (also called RAN1), plays a role in the transmembrane transport of Cu at the cytosol level to the Golgi apparatus (Proulx et al. 2017).

The HMA7/RAN1 (responsive to antagonist 1) protein is part of the P-type ATPase type P carrier (subtype 1B-1) (Hirayama et al. 1999). In *A. thaliana*, it is known that RAN1 is located on post-Golgi vesicles and is involved in the transport of Cu within this endomembrane system where this metal will be incorporated into ethylene receptors (Hirayama et al. 1999; Woeste and Kieber 2000; Yruela 2005). The gene coding for this protein has been associated with Cu resistance in *A. thaliana* (Kobayashi et al. 2007). Copper transporter protein (COPT 1) has been also associated with Cu resistance in *A. thaliana* (Sancenón et al. 2003; Proulx et al. 2017).

Metallothioneins (MTs) are cysteine-rich proteins found in the cytoplasm of cells of a wide variety of organisms such as animals, microorganisms, and plants (van Hoof et al. 2001). There is evidence that MTs play a role in protecting against the harmful effects of Cu through its chelation process (van Hoof et al. 2001; Guo et al. 2008). The chelation of Cu makes it unavailable for the formation of reactive species derived from oxygen (Guo et al. 2008). The implication of MTs in metal resistance is attributed to the presence of cysteine residues. These amino acids are predominantly present at the level of the units, whose function is to chelate metals such as Cd, Cu, and Zn (Guo et al. 2008).

More pertinently, several studies have demonstrated the role of MTs in multivariate tolerance (Cu, Cd, and Zn) in plant-based organisms (Ren and Zhao 2009; Lv et al. 2013). In plants, MTs are divided into four subgroups (MT1–MT4) as a function of cysteine residue distribution (Cobbett and Goldsbrough 2002). A particular subgroup, MT2, is an important determinant of Cu resistance in *A. thaliana*. A gene specific to the MT2 group has shown an increase in expression in several species in the presence of excess Cu. For example, copper treatment with *A. thaliana* increases the expression of MT2b in the roots (Guo et al. 2003). Other studies have shown that the populations of *Silene vulgaris* and *Silene paradoxa* (van Hoof et al. 2001; Mengoni et al. 2003) that are tolerant to Cu have a greater expression of the MT2b gene.

Proteins associated with multidrug resistance-associated proteins (MRPs) are proteins of the ABC carrier family (Rea 1999). Bovet et al. (2003) demonstrated an increase in expression of several genes encoding MRPs in response to Cd exposure in *A. thaliana*. In addition, Keinanen et al. (2007) compared Cu-resistant and Cu-sensitive birch (*Betula pendula*) and revealed differential expression of the MRP4 gene. They observed a significant increase in the expression of the MRP4 gene in Cu-resistant *B. pendula* compared to susceptible genotypes in the presence of a large amount of this metal (Keinanen et al. 2007).

Table 3 highlights genes involved in Cd resistance. Plants show varying degrees of resistance to this metal, but apart from the well-described metal tolerance mechanisms that have been already described, they have other proteins that participate in novel Cd tolerance mechanisms. A novel gene, *AtPcr1*, associated with Cd resistance in yeast and *Arabidopsis* was identified by Song et al. (2004). It mediates this resistance by reducing Cd uptake when expressed (Song et al. 2004). Another gene MuSI confers resistance to tobacco plants by reducing Cd uptake and/or increasing Cd immobilization in the roots resulting in less Cd translocation in the shoots (Kim et al. 2011). Zhang et al. (2016) have determined a series of 24 candidates Cd tolerant/ detoxification functional genes which could elucidate the molecular mechanisms of tobacco response to Cd. Chen et al. (2015) characterized a gene, MAN3, which regulates Cd tolerance through the glutathione-dependent pathway in *A. thaliana*.

2.4 Metal-Specific Induction of Gene Expression

Theriault et al. (2016a, b) reported the downregulation of AT2G16800, GR, and ZAT11 and the upregulation of IREG1 in *B. papyrifera* treated with high dose of Ni (1,600 mg/kg of soil). A downregulation of AT2G16800 was also observed in *Populus tremuloides* exposed to a high level of Ni (1,600 mg/kg) (Czajka et al. 2018). On the other hand, Djeukam and Nkongolo (2018) reported that Ni can trigger an increased expression of the targeted *ACC deaminase*, *SAT*, and *NAS3* at high doses in *Quercus rubra*. Proulx et al. (2017) reported that RAN1 and MT2b were downregulated in response to a high dose of Cu treatment (1,312 mg/kg) while MRP4 was upregulated. No expression of these genes was induced by a low dose of

Table 3 Candidate genes involved in cadmium resistance in model and non-model plant species

Gene	Species	Reference
Snakin-2, metallothionein 1, metallothionein 2, metallothionein 3, metallothionein 4, metallothionein 5, Rar1, copper transporter, PLAC8 family, GRP-2, basic PR, peptidase isoform 1, metallothionein-like, ubiquitin-like, calreticulin, CBP20, chitinase, GST1, GST2, PUB1, GST3, APX, and BPS1[a]	*Nicotiana tabacum* (L.)	Zhang et al. (2016)
AtPcr1[a]	*Arabidopsis* spp.	Song et al. (2004)
AtfC1 (cadmium stress response gene)	*Arabidopsis* spp.	Song et al. (2017)
MAN3 MuSI	*Arabidopsis thaliana* *Nicotiana tabacum* (L.)	Chen et al. (2015) Kim et al. (2011)

[a]These genes are also involved in yeast tolerance to cadmium

metals in these plant species (*B. papyrifera*, *P. tremuloides*, *Q. rubra*, and *A. rubrum*) (Theriault and Nkongolo 2016; Czajka et al. 2018; Djeukam et al. 2016; Kalubi et al. 2018). Konlechner et al. (2013) described some MTs and cell wall-modifying genes that were induced by Zn and Cd irrespective of the genotypes' origin and metal uptake capacity and a cysteine biosynthesis gene that was expressed constitutively higher in the metallicolous genotypes in willow (*Salix caprea* L.). Yu et al. (2017) identified genes that were upregulated in some pak choi (*Brassica rapa* subsp. *chinensis*) cultivars and downregulated in others when exposed to Cd.

2.4.1 Differential Gene Expression in Metal Resistant Plants

What distinguishes a resistant and susceptible genotype is the level of gene expression. Therefore, resistance is mainly due to changes in expressions from mutations in the promoter region or gene duplication (Freeman et al. 2005; Maron et al. 2013). High expression of transporters, chelators, or other genes involved in metal detoxifications is usually associated with metal tolerance. But, it is important to note that metal resistance is not only associated to high or low expression. If a mutation occurs in a gene and leads to the relocation or modification of the native protein, it could have a direct impact on metal resistance/sensitivity. Relocation of metal transporters can be achieved when there are mutations in the N-terminal of the gene, where signaling peptides can be found. For instance, deletion of nine amino acids from the N-terminus (putative signaling peptide) of the IRT1 transporter in *A. thaliana* increases Fe deficiency (Barberon et al. 2011).

3 Effects of Metals on Soil Microorganisms

Soil microbial community and their activities can be affected for decades by metal contamination. Toxic effects of metals on soil microorganisms depend on their bioavailability (Sherameti and Varma 2015; Winterhalder 1996; Gunn et al. 2007; Anand et al. 2003; Kamal et al. 2010). Significant reduction in a number of microbes, microbial growth rates, biomass, activity, community composition, and diversity have been observed in soils contaminated with metals (Sherameti and Varma 2015; Winterhalder 1996; Pennanen et al. 1998; Anand et al. 2003; Kamal et al. 2010). Metal contamination at low levels has adverse effects on bacteria but not on fungi (Pennanen et al. 1998; Kamal et al. 2010; Rajapaksha et al. 2004; Mühlbachová and Tlustoš 2006).

Soil microorganisms vary widely in their tolerance to metal contamination. For example, in bacteria, gram-negative cell walls have a lower charge capacity than gram-positive walls. However, they have a complex three-layered structure that binds and immobilizes many metal ions (Hg^{+} and Pb^{2+}) (Kamal et al. 2010; Hughes and Poole 1989; Mitchell and Gu 2009). Studies have shown that gram-negative bacteria exhibit greater metal tolerance than gram positive (Kamal et al. 2010; Hughes and Poole 1989; Mitchell and Gu 2009). Various studies have reported a decrease in bacterial activity but an increase in fungal activity in metal-contaminated soils (Sherameti and Varma 2015; Kamal et al. 2010; Rajapaksha et al. 2004; Mühlbachová and Tlustoš 2006). Other studies also reported a delay, reduced, and complete elimination of arbuscular mycorrhizal fungi (AM) and ectomycorrhizal fungi (ECM) colonization and spore germination in metal-contaminated soils (Kamal et al. 2010; Saraswat and Rai 2011; Bellion et al. 2006; Jentschke and Godbold 2000). In other studies, metal contamination had no significant effect on fungal development, which could be due to the presence of different ecotypes exhibiting different degrees of tolerance to metals (Kamal et al. 2010; Bellion et al. 2006; Jentschke and Godbold 2000). In polluted sites, fungi with higher tolerance to Cu, Zn, Cd, and Pb compared to those from unpolluted soils have been identified. Therefore, relatively high fungal colonization can be observed in plants growing in highly polluted soils (Kamal et al. 2010; Bellion et al. 2006; Jentschke and Godbold 2000). However, at the highest level of soil pollutions, fungal ecotypes and diversity diminish sharply due to toxic effect of metals causing an inability of certain fungal species to colonize root system and/or multiply in rhizosphere (Kamal et al. 2010; Bellion et al. 2006; Jentschke and Godbold 2000).

3.1 Metal Resistance in Microorganisms

There are significant similarities in metal resistance mechanisms between microorganisms and plants for components involved in metal chelation, sequestration, and transport. Metal resistance in microorganisms is usually due to metal chelation with organic ligands, sequestrations, transportation out of cells, and biotransformation of ions to less bioavailable or less toxic form (Kamal et al. 2010; Hughes and Poole 1989; Mitchell and Gu 2009). The resistance mechanism varies also with microbial species involved. Uranium (Ur) has been shown to rapidly accumulate in cells of *Saccharomyces cerevisiae* and *Pseudomonas aeruginosa* (Hughes and Poole 1989; Mitchell and Gu 2009). *Bacillus subtilis* cell walls bind to numerous metals [Hg^{2+}, Pb^{2+}, Mg^{2+}, Fe^{2+}, Cu^{2+}, Na^{+}, and potassium (K^{+})]. This binding is a result of the presence of many anionic sites on the cell wall, particularly (1) phosphodiester groups of teichoic acids, (2) carboxyl group of peptidoglycan, (3) sugar hydroxyl group of wall polymer, and (4) amide group of peptide chain (Hughes and Poole 1989; Mitchell and Gu 2009). Similarly, growing *S. cerevisiae* in mercuric chloride ($HgCl_2$) resulted in extensive metal binding to high affinity sites on the cell wall (Hughes and Poole 1989). Hyphal walls in *Neurospora crassa* bind to Cu, Fe, and Zn (Hughes and Poole 1989).

Some microorganisms (*Thiobacillus*, *Serratia*, *Pseudomonas*, *Bacillus*, *Penicillium*, *Aspergillus*) produce various organic and inorganic acids that can extract metals from solid substrates (Hughes and Poole 1989; Mitchell and Gu 2009). In addition to metal solubilization and transport, interaction between bacterial exopolysaccharides and metals has also been identified. Charged functional groups such as phosphate, carboxyl, and hydroxyl on the exopolymer are involved in adsorption of metals onto cell wall (Hughes and Poole 1989; Mitchell and Gu 2009). A number of bacterial and fungal species have the capability to transform metals to nontoxic form through methylation (Hughes and Poole 1989; Mitchell and Gu 2009).

3.2 Genetic Resistance to Metals in Soil Microorganisms

Genes for metal resistance in microorganisms are often present in plasmids and are easily disseminated through a population or community in response to selection pressure associated with metal exposure (Sherameti and Varma 2015; Kamal et al. 2010; Karelová et al. 2011; Manasi et al. 2016).

3.2.1 Genes for Metal Resistance in Bacteria and Archaea

In bacteria, efflux pumping is the basis of toxic ion resistance, involving transporters such as P-type ATPases or cation/H^{+} antiporter (Hall 2002). These pumping systems are involved in Cu, Cd, Zn, Co, and Ni resistance. Specific *czc* genes that encode for a cation-proton antiporter (CzcABC) which is responsible for resistance to Cd, Zn, and Co have been identified in gram-negative bacteria (e.g., *Ralstonia eutropha* and *Cupriavidus metallidurans*) (Kamal et al. 2010; Karelová et al. 2011). Similarly, *ncc* genes were found in gram-negative bacteria (e.g., *Alcaligenes xylosoxidans* also known as *Achromobacter xylosoxidans*) resistant to Ni, Cd, and Co. Cd resistance is linked to Cd-efflux ATPase and *cad* operon which have been identified in gram-positive bacteria (e.g., *Staphylococcus, Bacillus*, or *Listeria*) (Kamal et al. 2010; Karelová et al. 2011). Cu resistance has been associated with *cop* system (*Pseudomonas syringae*) and *pco* system (*E. coli*) where *cop* genes encode for different Cu-binding proteins that allow sequestration of Cu and where *pco* system is an ion-dependent Cu antiporter (Kamal et al. 2010).

Active efflux or trapping of the metal ions by metal chaperone is a typical mechanism used by bio-mining bacteria to resist high levels of Cu. Other strategies such as gene duplications, the presence of genomic islands, and other passive mechanisms have been also proposed (Navarro et al. 2013; Wen et al. 2014; Wheaton et al. 2015; Martínez-Bussenius et al. 2017). These mechanisms have been also identified in some archaea species (Orell et al. 2010). In bacteria, Rus and AcoP genes have been associated with Cu resistance in *Acidithiobacillus ferrivorans* and a putative MCO gene in *A. thiooxidans* (Valdes et al. 2011; González et al. 2014; Martínez-Bussenius et al. 2017). For archaea, a few metal efflux pumps have been described. In addition, genes encoding for a new type of transcriptional regulatory (CopT), a putative metal-binging chaperone (CopM), and putative Cu-transporting P-type ATPases have been identified (Wheaton et al. 2015; Martínez-Bussenius et al. 2017). Analysis of *Halobacterium* sp. Strain NRC-1 has shown that transporters such as putative Cd (II)-efflux ATPase could be involved in metal resistance (Kaur et al. 2006). Upregulation of yvgX, a P1B-type ATPase, in response to Cu and Zn contamination in this strain has been reported. Likewise, arsM has been associated to metal resistance in this species (Wang et al. 2004a, b). In comparison with plants, the metal tolerance proteins (MTP) were described as metal efflux transporters from the cytoplasm, transporting mainly Zn^{2+} but also Mn^{2+}, Fe^{2+}, Cd^{2+}, Co^{2+}, and Ni^{2+} (Ricachenevsky et al. 2013).

There are various metal-binding proteins involved in intracellular chelation which constitutes metal resistance in microorganisms. Some of these proteins include GSH and MT described in plants (Avery 2001). In haloarchaea, they have an equivalent protein to that of GSH. This protein is called γ-glutamylcysteine (γ-GC) (Sundquist and Fahey 1989; Malki et al. 2009), and it is responsible for coping with oxidative and disulfide stress in detoxification of metals (Fahey 2001). This protein's involvement with metal resistance has to do with its thiol group which is capable of chelating metals. Metal-chelating proteins are capable of undergoing multimerization with metals in archaea which leads to precipitation (Bini 2010). The metal-protein complex reverts back to its original state by redissolving itself, and the complex collapses once metal levels in the environment have weakened (Reindel et al. 2002; Tanaka et al. 2004). This process of multimerization is unique to archaea only. Another way for organisms to cope with metal stress in the environment is by the secretion of biofilm on the surface composed of exopolysaccharides (EPS) (Harrison et al. 2009; Gadd 2010). These organisms exhibit a different phenotype under these conditions which is based on transcription and growth rate (Donlan and Costerton 2002). The haloarchaea are capable of producing EPS as a stress response mechanism to protect themselves against nutrient deficiency, temperature fluctuation, and xenobiotics (Poli et al. 2011). The hyperthermophilic archaeon *Archaeoglobus fulgidus* has been reported to secrete this biofilm and to keep the metals trapped in the EPS matrix (Poli et al. 2011). Similar mechanisms have been observed in plants. It is possible that haloarchaea are able to do the same thing.

Studies have shown that many different operons exist in bacteria that confer metal resistance to specific types (Srivastava and Kowshik 2013). These metal-resistant determinants have been studied extensively in bacteria and all archaea except in haloarchaea (Ehrlich 1997; Dopson et al. 2003). In archaea, the arrangement of metal transporters as operons has not been shown except for ArsA ATPase transporter which is the ars operon that confers resistance against arsenic in *Halobacterium* sp. strain NRC-1 (Wang et al. 2004a, b). Haloarchaea have plasmids that are fairly large that contain genes for antibiotic resistance or metal resistance. These plasmids are called minichromosomes or megaplasmids (DasSarma et al. 2009). Some organisms have transporter genes downregulated for influx and upregulation of genes for efflux pumps. This will allow metals to be removed more efficiently (Gadd 2010). Haloarchaea harbor genes associated with influx and efflux transporters. These genes play an important role in metal resistance. Some of the transporters found in haloarchaea are P_{1B}-Type ATPases, cation diffusion facilitator (CDF) family, metal transporters, and ABC transporters (Srivastava and Kowshik 2013).

3.2.2 Genes for Metal Resistance in Fungi

For fungi, complexation of Cd by MT and GSH, for example, is a key mechanism for Cd tolerance in ECM fungus (*Paxillus involutus*) (Ramesh et al. 2009; Saraswat and Rai 2011). Two MTs have been identified in *S. cerevisiae*: one is induced by Cu (encoded by *CUP1*), and the second is regulated by Cu, Zn, and oxidative stress (encoded by *Crs5*) (Ramesh et al. 2009). MTs have been identified in a number of fungal species such as *Agaricus bisporus*, *Gigaspora margarita*, *Neurospora crassa*, *Pyrenopeziza brassicae*, *Podospora anserine*, and *P. involutus* (Bellion et al. 2006; Ramesh et al. 2009; Saraswat and Rai 2011). Two fungal species (*Candida glabrata* and *Schizosaccharomyces pombe*) produce both MT and PC (Saraswat and Rai 2011), while *C. glabrata* produce metal-chelating MT when exposed to toxic concentrations of Cu. But under Cd stress, it strictly produces PCs (Saraswat and Rai 2011). In addition to avoidance and compartmentalization, fungi also have other defense mechanisms that reduce oxidative stress generated by metal-induced ROS by synthesis of antioxidant enzymes such as superoxide dismutase (SOD), catalases (CATs), ascorbate peroxidase (APX), GSH, GST, and GR (Bellion et al. 2006; Saraswat and Rai 2011). As discussed in previous sections, the same mechanisms have been reported in plants. When grown on media with varying concentrations of Zn, *Glomus intraradices* was found to code for GST, SOD, cytochrome P450, and thioredoxin enzymes which are involved in detoxification of ROS (Saraswat and Rai 2011).

3.3 *Metal Specific Induction of Gene Expression in Microorganisms*

In the presence of metals, microorganisms like plants are required to adapt to environmental changes by triggering global transcriptional regulations. These regulations can either detoxify metals or prevent the metals from entering the cell. In most cases, transcription and translation are upregulated during early exposure to a stressor. This allows new proteins to be made and replace the ones that were damaged by the metals (Usadel et al. 2008; Mittal et al. 2012). Rademacher and Masepohl (2012) reviewed the literature on copper-responsive gene regulation in bacteria. They described multiple Cu-responsive regulators in different species. Only one study has been reported in haloarchaea on transcriptional changes to metal stress in *Halobacterium* sp. (Kaur et al. 2006). It is possible that changes in the transcriptome could have an impact on metal resistance in haloarchaea (Srivastava and Kowshik 2013).

3.4 *Cross-Organism Expression of Metal-Associated Genes in Plants and Microorganisms*

Horizontal gene transfer is a common phenomenon in species evolution. Transfer of genes associated with metal resistance from microorganisms to plants has not been documented in details. A recent analysis of gene expression in *Betula papyrifera* revealed that TonB-like protein associated with Ni resistance in this species is likely derived from a microorganism through horizontal transfer from prokaryotes (Theriault et al. 2016b). Functional expression of a bacterial metal transporter in *Arabidopsis* that enhances resistance to and decreases uptake of metals has been reported (Lee et al. 2003). A recent study by Zhang et al. (2016) showed that because plants are multicellular organism, translocation of metals such as Cd between different cells is mainly mediated by transporters or some secretory proteins. They showed that tobacco Cd tolerance genes increase tolerance to Cd in yeast, but the effects generated by different genes were not the same. In yeast, their detoxification effect is mainly mediated by chelation or cellular compartmentalization. Some genes cause Cd detoxification by secretion. In tobacco on the other hand, the acclimation to Cd stress is a complex process associated with broad gene functions.

4 Conclusion

There are significant similarities in metal resistance mechanisms between microorganisms and plants for components involved in metal chelation, sequestration, and transport. Because of plant complexity, the cellular responses to metal contamination in plants are quite different from those of microorganisms. Plant response to metal toxicity appears to be species-specific. Most of hyperaccumulators are from the *Brassicaceae* family, but variation in copying mechanism to metal contamination exists among species within the same genus. For example, silver maple (*Acer saccharinum*) is Ni excluder, while red maple (*Acer rubrum*) is Ni avoider. Knowledge of plant cellular mechanisms involved in metal toxicity is mostly from studies of model plants such as *Arabidopsis* spp. Transposition of these studies in non-model species resulted in different responses. Discovery of candidate genes associated to metal resistance in higher vascular plants is time consuming and cost prohibitive because of the complexity of their genomes and transcriptome. Recent discovery of genes associated directly with Ni, Cu, Cd, and other metal resistance in plants is promising and suggests that genes for metal tolerance could be used in the future to design metal-resistant plants/trees. On the other hand, significant progress has been made to better understand metal resistance in bacteria, archaea, and fungi. A number of microbial responses to metal contamination have been decrypted. Horizontal gene transfer between bacteria and archaea is a common phenomenon. The identification of horizontal transfers of genes from bacteria to *Betula* spp. is of paramount importance in shaping up the evolution of metal resistance in plants.

5 Summary

Toxicity to metals depends on the metal itself and its total/bioavailable concentration in soils. Metal toxicity in plants is not as studied as in other organisms such as microorganisms and animals. It is established that elevated levels of metals can stimulate in plants and microorganism formation of free radicals and reactive oxygen species (ROS) such as singlet oxygen (O_2), hydrogen peroxide (H_2O_2), and hydroxyl radicals ($-OH$) that cause oxidative stress resulting in cellular damage. The most recent review on metal resistance mechanisms in plants focused only on the role of transcriptomics, proteomics, metabolomics, and ionomics. Similarly, a comprehensive review of metal homeostasis and resistance in bacteria that mainly described transcriptional response to metal stress and posttranscriptional regulatory systems has been recently published. In this review, key relevant studies on strategies used by plants and microorganisms to cope with metals as well as the genetic mechanisms and genes associated with metals resistance are discussed. Plants use different approaches to cope with metals once they access cellular or intercellular compartments. They include restriction of metal uptake and transport, complexation of metals at the cell wall-plasma membrane interface, and complexation and compartmentation of metals with the cells. Classification of metal plant resistance mechanisms and genes involved in metal resistance such as chelators and transporters are also discussed. Interaction between metals and bacteria, archaea, and fungi as well as recent discoveries in genetic resistance mechanisms to metals in these microorganisms are described in details.

Acknowledgements This research is supported by the Natural Research Council of Canada (NSERC).

Conflict of Interest Authors declare no conflicts of interest.

References

Abdu N, Abdullahi AA, Abdulkadir A (2017) Heavy metals and soil microbes. Environ Chem Lett 15:65–84

Ahmad I, Hayat S, Pichtel J (2005) Heavy metal contamination of soil: problems and remedies. Science Publishers, Enfield

Alloway BJ (2014) Heavy metals in soils: trace metals and metalloids in soils and their bioavailability, 3rd edn. Springer, Dordrecht

Anand M, Ma K-M, Okonski A et al (2003) Characterising biocomplexity and soil microbial dynamics along a smelter-damaged landscape gradient. Sci Total Environ 311:247–259

Avery SV (2001) Metal toxicity in yeasts and the role of oxidative stress. Adv Appl Microbiol 49:111–142

Aydinalp C, Marinova S (2003) Distribution and forms of heavy metals in some agricultural soils. Pol J Environ Stud 12:629–633

Baloun J, Nevrtalova E, Kovacova V et al (2014) Characterization of the HMA7 gene and transcriptomic analysis of candidate genes for copper tolerance in two Silene vulgaris ecotypes. J Plant Physiol 171:1188–1196

Barberon M, Zelazny E, Robert S et al (2011) Monoubiquitin-dependent endocytosis of the iron-regulated transporter 1 (IRT1) transporter controls iron uptake in plants. Proc Natl Acad Sci 108:450–458

Bellion M, Courbot M, Jacob C et al (2006) Extracellular and cellular mechanisms sustaining metal tolerance in ectomycorrhizal fungi. FEMS Microbiol Lett 254:173–181

Bingham F, Pereyea F, Jarrell W (1986) Metal toxicity to agricultural crops. Met Ions Biol Syst 20:119–156

Bini E (2010) Archaeal transformation of metals in the environment. FEMS Microbiol Ecol 73:1–16

Bourg ACM (1995) Speciation of heavy metals in soils and groundwater and implications for their natural and provoked mobility. In: Förstner U, Salomons W, Mader P (eds) Heavy metals. Springer, Berlin, pp 19–31

Bovet L, Eggmann T, Meylan-Bettex M et al (2003) Transcript levels of AtMRPs after cadmium treatment: induction of AtMRP3. Plant Cell Environ 26:371–381

Burkhead JL, Gogolin Reynolds KA, Abdel-Ghany SE et al (2009) Copper homeostasis. New Phytol 182:799–816

Burton KW, Morgan E, Roig A (1983) The influence of heavy metals upon the growth of sitka-spruce in South Wales forests. Plant Soil 73:327–336

Chandrangsu P, Rensing C, Helmann JD (2017) Metal homeostasis and resistance in bacteria. Nat Rev Microbiol 15:338–350

Chen J, Yang L, Gu J et al (2015) MAN3 gene regulates cadmium tolerance through the glutathione-dependent pathway in Arabidopsis thaliana. New Phytol 205:570–582

Cobbett C, Goldsbrough P (2002) Phytochelatins and metallothioneins: roles in heavy metal detoxification and homeostasis. Annu Rev Plant Biol 53:159–182

Cobbett CS, May MJ, Howden R, Rolls B (1998) The glutathione deficient, cadmium sensitive mutant, cad2–1, of Arabidopsis thaliana is deficient in γ-glutamylcysteine synthetase. Plant J 16:73–78

Cuypers A, Remans T, Weyens N et al (2013) Soil-plant relationships of heavy metals and metalloids. In: Alloway BJ (ed) Heavy metals in soils. Springer, Dordrecht, pp 161–193
Czajka KM, Nkongolo K, Czajka KM, Nkongolo K (2018) High level of nicotianamine synthase (NAS3) and natural resistance associated macrophage protein (NRAMP4) gene transcription induced by potassium nitrate in trembling aspen (Populus tremuloides). Am J Biochem Biotechnol 14:183–190
Das R, Jayalekshmy VG (2015) Mechanism of heavy metal tolerance and improvement of tolerance in crop plants. J Glob Biosci 4:2678–2698
DasSarma S, Capes M, DasSarma P (2009) Haloarchaeal megaplasmids. In: Schwartz E (ed) Microbial megaplasmids. Springer, Berlin, pp 3–30
Deng X, He J, He N (2013) Comparative study on Ni2+-affinity transport of nickel/cobalt permeases (NiCoTs) and the potential of recombinant Escherichia coli for Ni2+ bioaccumulation. Bioresour Technol 130:69–74
Dixon RK, Buschena CA (1988) Response of ectomycorrhizal Pinus banksiana and Picea glauca to heavy metals in soil. Plant Soil 105:265–271
Djeukam CL, Nkongolo K (2018) Expression of genes associated with nickel resistance in red oak (Quercus rubra) populations from a metal contaminated region. Bull Environ Contam Toxicol 100:792–797
Djeukam CL, Theriault G, Michael P, Nkongolo KK (2016) Analysis of gene expression associated with copper toxicity in white birch (Betula papyrifera) populations from a mining region. Br Biotechnol J 15:1–10
Donlan RM, Costerton JW (2002) Biofilms: survival mechanisms of clinically relevant microorganisms. Clin Microbiol Rev 15:167–193
Dopson M, Baker-Austin C, Koppineedi PR, Bond PL (2003) Growth in sulfidic mineral environments: metal resistance mechanisms in acidophilic micro-organisms. Microbiology 149:1959–1970
Ehrlich HL (1997) Microbes and metals. Appl Microbiol Biotechnol 48:687–692. https://doi.org/10.1007/s002530051116
Emamverdian A, Ding Y, Mokhberdoran F, Xie Y (2015) Heavy metal stress and some mechanisms of plant defense response. Sci World J 2015:1–18
Fahey RC (2001) Novel thiols of prokaryotes. Annu Rev Microbiol 55:333–356
Foy CD, Chaney RL, White MC (1978) The physiology of metal toxicity in plants. Annu Rev Plant Physiol 29:511–566
Freeman JL, Persans MW, Nieman K et al (2004) Increased glutathione biosynthesis plays a role in nickel tolerance in thlaspi nickel hyperaccumulators. Plant Cell 16:2176–2191
Freeman JL, Garcia D, Kim D et al (2005) Constitutively elevated salicylic acid signals glutathione-mediated nickel tolerance in thlaspi nickel hyperaccumulators. Plant Physiol 137:1082–1091
Gadd GM (2010) Metals, minerals and microbes: geomicrobiology and bioremediation. Microbiology 156:609–643
Gajewska E, Skłodowska M (2008) Differential biochemical responses of wheat shoots and roots to nickel stress: antioxidative reactions and proline accumulation. Plant Growth Regul 54:179–188
González C, Yanquepe M, Cardenas JP et al (2014) Genetic variability of psychrotolerant Acidithiobacillus ferrivorans revealed by (meta)genomic analysis. Res Microbiol 165:726–734
Grill E, Löffler S, Winnacker E-L, Zenk MH (1989) Phytochelatins, the heavy-metal-binding peptides of plants, are synthesized from glutathione by a specific γ-glutamylcysteine dipeptidyl transpeptidase (phytochelatin synthase). Proc Natl Acad Sci 86:6838–6842
Gunn JM, Beckett PJ, Lautenback WE, Monet S (2007) Sudbury, Canada: from pollution record holder to award winning restoration site. In: France RL (ed) Handbook of regenerative landscape design. CRC Press, Boca Raton, pp 381–405
Guo W, Bundithya W, Goldsbrough PB (2003) Characterization of the Arabidopsis metallothionein gene family: tissue-specific expression and induction during senescence and in response to copper. New Phytol 159:369–381

Guo WJ, Meetam M, Goldsbrough PB (2008) Examining the specific contributions of individual Arabidopsis metallothioneins to copper distribution and metal tolerance. Plant Physiol 146:1697–1706
Hall JL (2002) Cellular mechanisms for heavy metal detoxification and tolerance. J Exp Bot 53:1–11
Harrison JJ, Turner RJ, Ceri H (2009) Metal tolerance in bacterial biofilms. Recent Res Dev Microbiol 9:33–55
Hasegawa I, Terada E, Sunairi M et al (1997) Genetic improvement of heavy metal tolerance in plants by transfer of the yeast metallothionein gene (CUP1). Plant Soil 196:227–281
Hassan Z, Aarts MGM (2011) Opportunities and feasibilities for biotechnological improvement of Zn, Cd or Ni tolerance and accumulation in plants. Environ Exp Bot 72:53–63
Henry JR (2000) An overview of the phytoremediation of lead and mercury. EPA, Washington
Hirayama T, Kieber JJ, Hirayama N et al (1999) RESPONSIVE-TO-ANTAGONIST1, a Menkes/Wilson disease-related copper transporter, is required for ethylene signaling in Arabidopsis. Cell 97:383–393
Hohl H, Varma A (2010) Soil: the living matrix. In: Sherameti I, Varma A (eds) Soil heavy metals. Springer, Berlin, pp 1–18
Hughes MN, Poole RK (1989) Metals and microorganisms. Chapman and Hall, London
Jentschke G, Godbold DL (2000) Metal toxicity and ectomycorrhizas. Physiol Plant 109:107–116
Kalubi KN, Michael P, Omri A (2018) Analysis of gene expression in red maple (Acer rubrum) and trembling aspen (Populus tremuloides) populations from a mining region. Genes Genom 40:1127–1136
Kamal S, Prasad R, Varma A (2010) Soil microbial diversity in relation to heavy metals. In: Sherameti I, Varma A (eds) Soil heavy metals. Springer, Berlin, pp 31–63
Karelová E, Harichová J, Stojnev T et al (2011) The isolation of heavy-metal resistant culturable bacteria and resistance determinants from a heavy-metal-contaminated site. Biologia 66:18–26
Kaur A, Pan M, Meislin M et al (2006) A systems view of haloarchaeal strategies to withstand stress from transition metals. Genome Res 16:841–854
Keinanen SI, Hassinen VH, Karenlampi SO, Tervahauta AI (2007) Isolation of genes up-regulated by copper in a copper-tolerant birch (Betula pendula) clone. Tree Physiol 27:1243–1252
Kim H, Seo Y (2012) Identification of potential molecular biomarkers in response to thioredoxin reductase 1 deficiency under nickel exposure. Biochip J 6:157–164
Kim S, Takahashi M, Higuchi K et al (2005) Increased nicotianamine biosynthesis confers enhanced tolerance of high levels of metals, in particular nickel, to plants. Plant Cell Physiol 46:1809–1818
Kim Y-N, Kim J-S, Seo S-G et al (2011) Cadmium resistance in tobacco plants expressing the MuSI gene. Plant Biotechnol 5:323–329
Kobayashi Y, Hoekenga OA, Itoh H et al (2007) Characterization of AtALMT1 expression in aluminum-inducible malate release and its role for rhizotoxic stress tolerance in Arabidopsis. Plant Physiol 145:843–852
Kobayashi Y, Kuroda K, Kimura K et al (2008) Amino acid polymorphisms in strictly conserved domains of a P-type ATPase HMA5 are involved in the mechanism of copper tolerance variation in Arabidopsis. Plant Physiol 148:969–980
Konlechner C, Türktaş M, Langer I et al (2013) Expression of zinc and cadmium responsive genes in leaves of willow (Salix caprea L.) genotypes with different accumulation characteristics. Environ Pollut 178:121–127
Krämer U, Cotter-Howells JD, Charnock JM et al (1996) Free histidine as a metal chelator in plants that accumulate nickel. Nature 379:635–638
Kushwaha A, Rani R, Kumar S, Gautam A (2016) Heavy metal detoxification and tolerance mechanisms in plants: implications for phytoremediation. Environ Rev 24:39–51
Lee J, Reeves RD, Brooks RR, Jaffré T (1977) Isolation and identification of a citrato-complex of nickel from nickel-accumulating plants. Phytochemistry 16:1503–1505

Lee J, Bae H, Jeong J et al (2003) Functional expression of a bacterial heavy metal transporter in Arabidopsis enhances resistance to and decreases uptake of heavy metals. Plant Physiol 133:589–596

Lee J, Donghwan S, Won-yong S et al (2004) Arabidopsis metallothioneins 2a and 3 enhance resistance to cadmium when expressed in Vicia faba guard cells. Plant Mol Biol 54:805–815

Lemaire S, Guillon B, Le Marechal P et al (2004) New thioredoxin targets in the unicellular photosynthetic eukaryote Chlamydomonas reinhardtii. Proc Natl Acad Sci U S A 101:7475–7480

Liu J, Magalhaes JV, Shaff J, Kochian LV (2009) Aluminum activated citrate and malate transporters from the MATE and ALMT families function independently to confer Arabidopsis aluminum tolerance. Plant J 57:389–399

Liu X-M, An J, Han H et al (2014) ZAT11, a zinc finger transcription factor, is a negative regulator of nickel ion tolerance in Arabidopsis. Plant Cell Rep 33:2015–2021

Lv Y, Deng X, Quan L et al (2013) Metallothioneins BcMT1 and BcMT2 from Brassica campestris enhance tolerance to cadmium and copper and decrease production of reactive oxygen species in Arabidopsis thaliana. Plant Soil 367:507–519

Makela M, Michael P, Theriault G, Nkongolo KK (2016) High genetic variation among closely related red oak (Quercus rubra) populations in an ecosystem under metal stress: analysis of gene regulation. Genes Genom 38:967–976

Malki L, Yanku M, Borovok I et al (2009) Identification and characterization of gshA, a gene encoding the glutamate-cysteine ligase in the halophilic archaeon Haloferax volcanii. J Bacteriol 191:5196–5204

Manasi, Rajesh N, Rajesh V (2016) Evaluation of the genetic basis of heavy metal resistance in an isolate from electronic industry effluent. J Genet Eng Biotechnol 14:177–180

Mari S, Gendre D, Pianelli K et al (2006) Root-to-shoot long-distance circulation of nicotianamine and nicotianamine-nickel chelates in the metal hyperaccumulator Thlaspi caerulescens. J Exp Bot 57:4111–4122

Maron LG, Guimaraes CT, Kirst M et al (2013) Aluminum tolerance in maize is associated with higher MATE1 gene copy number. Proc Natl Acad Sci U S A 110:5241–5246

Martínez-Bussenius C, Navarro CA, Jerez CA (2017) Microbial copper resistance: importance in biohydrometallurgy. Microb Biotechnol 10:279–295

McGrath SP, Chaudri AM, Giller KE (1995) Long-term effects of metals in sewage sludge on soils, microorganisms and plants. J Ind Microbiol 14:94–104

Mehes-Smith M, Nkongolo KK (2015) Physiological and cytological responses of Deschampsia cespitosa and Populus tremuloides to soil metal contamination. Water Air Soil Pollut 226:1–12

Mehes-Smith M, Nkongolo K, Cholewa E (2013) Coping mechanisms of plants to metal contaminated soil. In: Silvern S, Young S (eds) Environmental change and sustainability. InTech, Rijeka, pp 1–39

Mehrotra VS (2005) Mycorrhiza: role and applications. Allied Publishers, New Delhi

Mengoni A, Gonnelli C, Hakvoort HWJ et al (2003) Evolution of copper-tolerance and increased expression of a 2b-type metallothionein gene in Silene paradoxa L. populations. Plant Soil 257:451–457

Mitchell R, Gu J-D (2009) Environmental microbiology. Wiley, Hoboken

Mittal D, Madhyastha DA, Grover A (2012) Genome-wide transcriptional profiles during temperature and oxidative stress reveal coordinated expression patterns and overlapping regulons in rice. PLoS One 7:e40899

Mizuno T, Usui K, Horie K et al (2005) Cloning of three ZIP/Nramp transporter genes from a Ni hyperaccumulator plant Thlaspi japonicum and their Ni 2+-transport abilities. Plant Physiol Biochem 43:793–801

Mühlbachová G, Tlustoš P (2006) Effects of liming on the microbial biomass and its activities in soils long-term contaminated by toxic elements. Plant Soil Environ 52:345–352

Mulligan CN, Yong RN, Gibbs BF (2001) Remediation technologies for metal-contaminated soils and groundwater: an evaluation. Eng Geol 60:193–207

Navarro CA, von Bernath D, Jerez CA (2013) Heavy metal resistance strategies of acidophilic bacteria and their acquisition: importance for biomining and bioremediation. Biol Res 46:363–371
Orell A, Navarro CA, Arancibia R et al (2010) Life in blue: copper resistance mechanisms of bacteria and Archaea used in industrial biomining of minerals. Biotechnol Adv 28:839–848
Pennanen T, Fritze H, Vanhala P, Kiikkilä O (1998) Structure of a microbial community in soil after prolonged addition of low levels of simulated acid rain. Appl Environ Microbiol 64:2173–2180
Poli A, Di Donato P, Abbamondi GR, Nicolaus B (2011) Synthesis, production, and biotechnological applications of exopolysaccharides and polyhydroxyalkanoates by archaea. Archaea 2011:1–13
Prasad MNV (2014) Metallothioneins, metal binding complexes and metal sequestration in plants. In: Heavy metal stress in plants. Springer, Berlin, pp 47–83
Proulx M, Michael P, Djeukam C, Nkongolo K (2017) Differential gene transcription in red oak (Quercus rubra) genotypes resistant to copper toxicity. Am J Biochem Biotechnol 13:215–225
Rademacher C, Masepohl B (2012) Copper-responsive gene regulation in bacteria. Microbiology 158:2451–2464
Rajapaksha RMCP, Bååth E, Ba E (2004) Metal toxicity affects fungal and bacterial activities in soil differently metal toxicity affects fungal and bacterial activities in soil differently. Appl Environ Microbiol 70:2966–2973
Ramesh G, Podila GK, Gay G et al (2009) Different patterns of regulation for the copper and cadmium metallothioneins of the ectomycorrhizal fungus Hebeloma cylindrosporum. Appl Environ Microbiol 75:2266–2274
Rauser WE, Curvetto NR (1980) Metallothionein occurs in roots of Agrostis tolerant to excess copper. Nature 287:563–564
Rea PA (1999) MRP subfamily ABC transporters from plants and yeast. J Exp Bot 50:895–913
Reindel S, Anemüller S, Sawaryn A, Matzanke BF (2002) The DpsA-homologue of the archaeon Halobacterium salinarum is a ferritin. Biochim Biophys Acta 1598:140–146
Ren Y, Zhao J (2009) Functional analysis of the rice metallothionein gene OsMT2b promoter in transgenic Arabidopsis plants and rice germinated embryos. Plant Sci 176:528–538
Ricachenevsky FK, Menguer PK, Sperotto RA et al (2013) Roles of plant metal tolerance proteins (MTP) in metal storage and potential use in biofortification strategies. Front Plant Sci 4:144
Rodionov DA, Hebbeln P, Gelfand MS, Eitinger T (2006) Comparative and functional genomic analysis of prokaryotic nickel and cobalt uptake transporters: evidence for a novel group of ATP-binding cassette transporters. J Bacteriol 188:317–327
Ryan PR, Raman H, Gupta S et al (2009) A second mechanism for aluminum resistance in wheat relies on the constitutive efflux of citrate from roots. Plant Physiol 149:340–351
Sancenón V, Puig S, Mira H et al (2003) Identification of a copper transporter family in Arabidopsis thaliana. Plant Mol Biol 51:577–587
Sancenon V, Puig S, Mateu-Andres I et al (2004) The Arabidopsis copper transporter COPT1 functions in root elongation and pollen development. J Biol Chem 279:15348–15355
Saraswat S, Rai JPN (2011) Mechanism of metal tolerance and detoxification in mycorrhizal fungi. In: Khan MS, Zaidi A, Goel R, Musarrat J (eds) Biomanagement of metal-contaminated soils. Springer, Dordrecht, pp 225–240
Schaaf G, Honsbein A, Meda AR et al (2006) AtIREG2 encodes a tonoplast transport protein involved in iron-dependent nickel detoxification in Arabidopsis thaliana roots. J Biol Chem 281:25532–25540
Schultz CL, Hutchinson TC (1988) Evidence against a key role for metallothionein like protein in the copper tolerance mechanism of Deschampsia cespitosa (L.) Beauv. New Phytol 110:163–171
Sharma SS, Dietz K-J (2009) The relationship between metal toxicity and cellular redox imbalance. Trends Plant Sci 14:43–50
Sherameti I, Varma A (2015) Heavy metal contamination of soils: monitoring and remediation. Springer, New York

Shojima S, Nishizawa N-K, Fushiya S et al (1990) Biosynthesis of phytosiderophores in vitro biosynthesis of 2′-deoxymugineic acid from L-methionine and nicotianamine. Plant Physiol 93:1497–1503
Singh S, Parihar P, Singh R et al (2015) Heavy metal tolerance in plants: role of transcriptomics, proteomics, metabolomics, and ionomics. Front Plant Sci 6:1143
Song W-Y, Martinoia E, Lee J et al (2004) A novel family of cys-rich membrane proteins mediates cadmium resistance in Arabidopsis. Plant Physiol 135:1027–1039
Song J, Feng SJ, Chen J et al. (2017) A cadmium stress-responsive gene AtFC1 confers plant tolerance to cadmium toxicity. BMC Plant Biol 17:187. https://doi.org/10.1186/s12870-017-1141-0
Srivastava P, Kowshik M (2013) Mechanisms of metal resistance and homeostasis in haloarchaea. Archaea 2013:1–16
Stearns JC, Shah S, Greenberg BM et al (2005) Tolerance of transgenic canola expressing 1-aminocyclopropane-1-carboxylic acid deaminase to growth inhibition by nickel. Plant Physiol Biochem 43:701–708
Sundquist AR, Fahey RC (1989) The function of gamma-glutamylcysteine and bis-gamma-glutamylcysteine reductase in Halobacterium halobium. J Biol Chem 264:719–725
Tanaka Y, Tsumoto K, Nakanishi T et al (2004) Structural implications for heavy metal-induced reversible assembly and aggregation of a protein: the case of Pyrococcus horikoshii CutA. FEBS Lett 556:167–174
Theriault G, Nkongolo K (2016) Nickel and copper toxicity and plant response mechanisms in white birch (Betula papyrifera). Bull Environ Contam Toxicol 97:171–176
Theriault G, Nkongolo KK (2017) Evidence of prokaryote like protein associated with nickel resistance in higher plants: horizontal transfer of TonB-dependent receptor/protein in Betula genus or de novo mechanisms? Heredity (Edinb) 118:358–365
Theriault G, Michael P, Nkongolo K (2016a) Decrypting the regulation and mechanism of nickel resistance in white birch (Betula papyrifera) using cross-species metal-resistance genes. Genes Genom 38:341–350
Theriault G, Michael P, Nkongolo K (2016b) Comprehensive transcriptome analysis of response to nickel stress in white birch (Betula papyrifera). PLoS One 11:1–20
Tong Y-P, Kneer R, Zhu Y-G (2004) Vacuolar compartmentalization: a second-generation approach to engineering plants for phytoremediation. Trends Plant Sci 9:7–9
Usadel B, Blasing OE, Gibon Y et al (2008) Global transcript levels respond to small changes of the carbon status during progressive exhaustion of carbohydrates in Arabidopsis rosettes. Plant Physiol 146:1834–1861
USDA NRCS (2000) Heavy metal soil contamination. Soil quality – urban technical notice 3. USDA NRCS, Auburn
Valdes J, Ossandon F, Quatrini R et al (2011) Draft genome sequence of the extremely acidophilic biomining bacterium Acidithiobacillus thiooxidans ATCC 19377 provides insights into the evolution of the Acidithiobacillus genus. J Bacteriol 193:7003–7004
van de Mortel JE, Villanueva LA, Schat H et al (2006) Large expression differences in genes for iron and zinc homeostasis, stress response, and lignin biosynthesis distinguish roots of Arabidopsis thaliana and the related metal hyperaccumulator Thlaspi caerulescens. Plant Physiol 142:1127–1147
van der Zaal B, Neuteboom L, Pinas J et al (1999) Overexpression of a novel Arabidopsis gene related to putative zinc-transporter genes from animals can lead to enhanced zinc resistance and accumulation. Plant Physiol 119:1047–1055
van Hoof NALM, Hassinen VH, Hakvoort HWJ et al (2001) Enhanced copper tolerance in Silene vulgaris (Moench) Garcke populations from copper mines is associated with increased transcript levels of a 2b-type metallothionein gene. Plant Physiol 126:1519–1526
Vert G, Grotz N, Dédaldéchamp F et al (2002) IRT1, an Arabidopsis transporter essential for iron uptake from the soil and for plant growth. Plant Cell 14:1223–1233
Viehweger K (2014) How plants cope with heavy metals. Bot Stud 55:1–12

Visioli G, Vincenzi S, Marmiroli M, Marmiroli N (2012) Correlation between phenotype and proteome in the Ni hyperaccumulator Noccaea caerulescens subsp. caerulescens. Environ Exp Bot 77:156–164

Wang G, Kennedy SP, Fasiludeen S et al (2004a) Arsenic resistance in Halobacterium sp. strain NRC-1 examined by using an improved gene knockout system. J Bacteriol 186:3187–3194

Wang W, Vinocur B, Shoseyov O, Altman A (2004b) Role of plant heat-shock proteins and molecular chaperones in the abiotic stress response. Trends Plant Sci 9:244–252

Wang D, Wei Z, Yang C, Liu G (2008) Analysis and identification of SCAR molecular markers associated with birch fiber length trait. J For Res 19:288–292

Wen Q, Liu X, Wang H, Lin J (2014) A versatile and efficient markerless gene disruption system for *Acidithiobacillus thiooxidans*: application for characterizing a copper tolerance related multicopper oxidase gene. Environ Microbiol 16:3499–3514

Wheaton G, Counts J, Mukherjee A et al (2015) The confluence of heavy metal biooxidation and heavy metal resistance: implications for bioleaching by extreme thermoacidophiles. Fortschr Mineral 5:397–451

Williams LE, Pittman JK, Hall JL (2000) Emerging mechanisms for heavy metal transport in plants. Biochim Biophys Acta Biomembr 1465:104–126

Winterhalder K (1996) Environmental degradation and rehabilitation of the landscape around Sudbury, a major mining and smelting area. Environ Rev 4:185–224

Woeste KE, Kieber JJ (2000) A strong loss-of-function mutation in RAN1 results in constitutive activation of the ethylene response pathway as well as a rosette-lethal phenotype. Plant Cell 12:443–455

Yadav SK (2010) Heavy metals toxicity in plants: an overview on the role of glutathione and phytochelatins in heavy metal stress tolerance of plants. S Afr J Bot 76:167–179

Yruela I (2005) Copper in plants. Braz J Plant Physiol 17:145–156

Yu R, Tang Y, Liu C et al (2017) Comparative transcriptomic analysis reveals the roles of ROS scavenging genes in response to cadmium in two pak choi cultivars. Sci Rep 7:9217

Zhang M, Mo H, Sun W et al (2016) Systematic isolation and characterization of cadmium tolerant genes in tobacco: a cDNA library construction and screening approach. PLoS One 11:e0161147

Zhu W, Zhao D-X, Miao Q et al (2009) Arabidopsis thaliana metallothionein, AtMT2a, mediates ROS balance during oxidative stress. J Plant Biol 52:585–592

Heavy Metal Threats to Plants and Soil Life in Southern Africa: Present Knowledge and Consequences for Ecological Risk Assessment

Herman Eijsackers, Adriaan Reinecke, Sophié Reinecke, and Mark Maboeta

Contents

H. Eijsackers
Unit for Environmental Sciences and Management, North-West University, Potchefstroom, South Africa

Wageningen University & Research, Wageningen, The Netherlands
e-mail: Herman.Eijsackers@wur.nl

A. Reinecke · S. Reinecke
Department of Botany and Zoology, Stellenbosch University, Matieland, South Africa
e-mail: ajr@sun.ac.za; sar@sun.ac.za

M. Maboeta (✉)
Unit for Environmental Sciences and Management, North-West University, Potchefstroom, South Africa
e-mail: mark.maboeta@nwu.ac.za

P. de Voogt (ed.), *Reviews of Environmental Contamination and Toxicology, Volume 249*, Reviews of Environmental Contamination and Toxicology Volume 249, https://doi.org/10.1007/398_2019_23

Abbreviations

AM	Arbuscular mycorrhiza
BAF	Bioaccumulation factor
CEC	Cation exchange capacity
DNA	Deoxyribonucleic acid
ERA	Ecological risk assessment
HM	Heavy metal
ISO	International Organization for Standardization
MTT	3-(4,5-Dimethylthiazol-2-Yl)-2,5-diphenyltetrazolium bromide spectrophotometric test
NRR	Neutral red spectrophotometric assay
NRRT	Neutral red retention test
OECD	Organisation for Economic Co-operation and Development
PAH	Poly aromatic hydrocarbons
RA	Risk assessment
SA	South Africa
SADC	Southern African Development Community
SSD	Species sensitivity distribution
SSV	Soil screening value
TDF	Tailing disposal facility
UNEP	United Nations Environment Programme
USA	United States

1 Introduction

Soil ecotoxicology and (eco)toxicological risk assessment have developed remarkably in recent times. From an addendum to human toxicology (Truhaut 1975), it has grown into a fully developed scientific discipline with its own unique theories and models. Nevertheless, it is debatable whether the presently developed risk assessment (RA) models for human and environmental exposures and effects are really representative of contaminated sites in all parts of the world. An ecological risk assessment (ERA) is used to evaluate the possible impact of environmental stressors on the environment. It will typically entail a three-step procedure including identification of a possible threat, risk characterization and risk evaluation. Laboratory testing, field surveys and risk assessment models are all utilized in ERA frameworks. The species sensitivity distribution (SSD) approach has been developed in North America and Europe (Posthuma et al. 2002) to address the enormous biodiversity in species. It is based on the assumption that a statistical sensitivity distribution of a "representative" number of (test) species from different taxa covers the sensitivities of all other species (Van Straalen and van Leeuwen 2002). However, the test species on which these distributions are based only cover a limited number of taxonomic phyla. Environmental conditions used in this approach are based on a combination of mean values within a statistically defined range. But the question is whether this will also be relevant for situations with a combination of extreme conditions and completely different species? Therefore, we need to know and ascertain how sensitive these different species are and to what extent different and more extreme environmental conditions influence toxicity of various contaminants. Southern African soils are characterized by a species composition that is for a considerable part endemic and in terms of environmental conditions considered to be extreme compared to global mean conditions (Eijsackers et al. 2017). Moreover, we have to realize that due to climate change effects, these conditions are expected to become more extreme, with more erratic changes and with resultant severe adverse impacts on biodiversity (Erasmus et al. 2002). Also, as a further consequence of climate change, contaminants could become more bioavailable which could then reinforce adverse impacts on natural life.

This paper intends to provide an inventory and overview of various research studies carried out in southern African countries. Further, to discuss this with respect to the prevailing environmental conditions, the occurring soil life and the diverse ways heavy metal (HM) contaminants are influencing it. To this end, we broadly describe southern African soils, soil life and living conditions. Next, we summarize how much research has been published on the topics mentioned above, originating from these southern African countries, and discuss the extent these results are suitable for application in standard risk assessment (RA) models currently employed in other countries of the western world. The aim is to determine whether an adapted approach is needed for this region.

Southern Africa is defined as the countries that constitute the Southern African Development Community (SADC). These are Angola, Botswana, the Democratic Republic of Congo, Lesotho, Madagascar, Malawi, Mauritius, Mozambique, Namibia, South Africa, Seychelles, Swaziland, United Republic of Tanzania, Zambia and Zimbabwe. Given the limited availability of literature for several of these countries, we confined our research primarily to Botswana, Namibia, South Africa, Zambia and Zimbabwe.

2 Methodology and Approach

2.1 Search Profile on Soil Contamination Related Research in Southern African Countries

To inventory the published literature on heavy metal impacts on plant and soil life with CAB abstracts, an analysis was performed using "Southern Africa explode". This covers Mozambique, Zimbabwe, Zambia, Botswana, Namibia, Angola, South Africa, Lesotho, Saint Helena, Swaziland and the Comoros (Fig. 1). A search profile was used with the terms "Soil? AND (contamin* or pollut* or heavy metals)

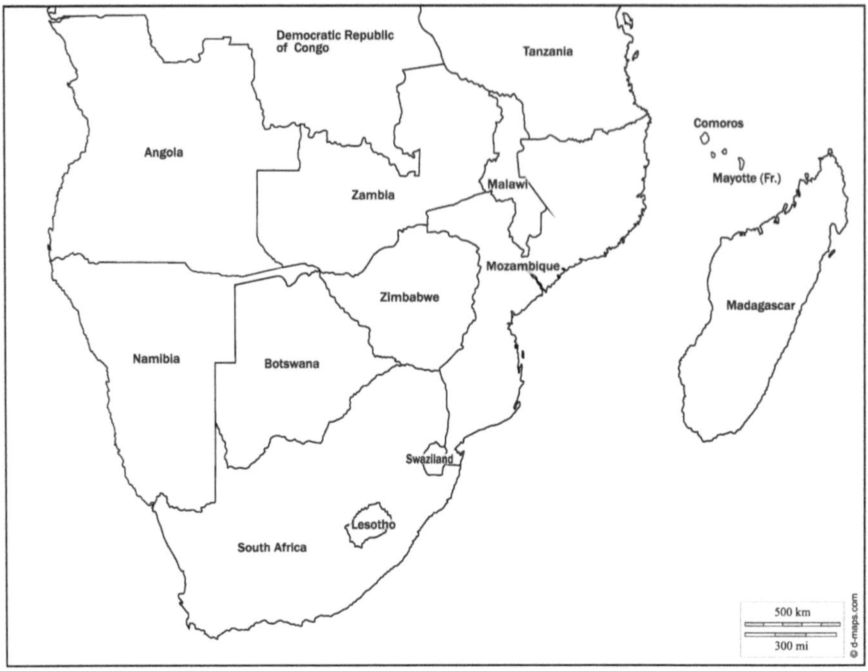

Fig. 1 Map of southern Africa (https://d-maps.com/carte.php?num_car=4359&lang=en). Mauritius (±940 km east of Nosy-Varika, Madagascar) and Seychelles (±1,730 km east of Mombasa, Kenya) not indicated on map

AND (Southern Africa explode)" resulting in 566 publications, of which 220 were from 2012 to 2018 or later, with 138 papers on contamination by inorganic mostly heavy metal compounds, of which 93 dealt with ecotoxicological impacts.

The 102 papers selected for further description were published in 2004 (6), 2005 (7), 2006 (14), 2007 (7), 2008 (2), 2009 (6), 2010 (9), 2011 (15), 2012 (10), 2013 (15) and 2014 (11). Although there is a gradual progress across the years, this is still limited. Further analyses of these papers indicate the following main themes with number of papers indicated in brackets: reference and background metal levels (13), biomarker studies (26), microbes and microbial interactions with plants (11), micro-fauna (45) and plants (39). Because quite a number of papers can be classified under various headings, the total number is higher than the number of papers (102) mentioned.

For the soil life, soil biomarkers and soil fauna sections, we also used the extensive datasets by A.J. Reinecke, S.A. Reinecke and M.S. Maboeta, bearing in mind their long history of researches in these fields as well as additional literature searches by these co-authors.

The general picture depicted from this information is that no recent research had been conducted focusing exclusively on soil policy in southern African states, especially research related to soil degradation due to contamination as already concluded by Eijsackers et al. (2006).

Most of the articles regarding soil contamination are on impacts of mining activities:

- Soil reference or background levels have been investigated for mining areas and tailing dams and some for roadsides or disposal sites, but there are very few real reference studies providing an overview of an area as a whole comparison to the noninfluenced natural area.
- Most plant research papers are on ornamental plants, with fewer publications on edible/medicinal plants for human use. A few papers addressed ecotoxicological impacts on plants or are about hyperaccumulators, i.e. plant species that strongly accumulate heavy metals and are sometimes applied to remediate contaminated sites.
- Research on soil animals has been carried out on different levels of biological organization (biochemical, physiological, cellular, organismal and population level), mostly reflecting laboratory experiments and only limited field research.
- Research on soil microflora is limited. A few papers report relations between plants and vesicular-arbuscular mycorrhiza (VAM) and impacts on the sensitivity of plants for heavy metals.
- Our literature search further showed that physical threats like erosion and salinization received much attention. In a preliminary search of 58 papers on soil degradation in relation to policy, 56 dealt with these aspects either directly or related to overgrazing and improper land use. In the second search mentioned above, excluding the policy aspects, 410 papers were collected, of which 165 dealt with different aspects of soil contamination.

2.2 *Soils and Soil Degradation and Legislation in Southern African Countries*

Most of the soils in southern Africa are poor and dry, with the exception of the tropical areas in the north-eastern part. The soils in the studied area are often maltreated and mismanaged without considerations of proper soil management principles. Many of the soils have limited arability which currently has an impact on food security, as well as in the future (Fey 2010; Mugerwa 2015).

On the Highveld regions, strong winds and water and wind erosion vary considerably from east to west and north to south. Deep residual soils are usually found in areas of relatively high rainfall in the eastern areas of southern Africa, and chemical weathering is dominant. Shallow bedrock with a thin cover of residual soil is usually found in the arid western areas of southern Africa (Webster and Fey 2011).

Climatic and physical factors such as rainfall, soil type and variable topography, combined with certain agricultural practices, constitute the most important immediate causes of the degradation of soils. Moreover, there are threats due to mining (Rösner and Van Schalkwyk 2000), various types of waste deposits (Reinecke et al. 2016) and large-scale application of pesticides (Dalvie et al. 2009).

There are also indirect impacts. Firstly, land use change is almost always combined with a change in the presence, mobility and activity of chemical compounds. Secondly climate change induces changed behaviour of soil organisms and changed exposure to and sensitivity for negative impacts of chemical contaminants. And thirdly there are indirect impacts due to biological invasions. The competition with introduced alien species may make indigenous species more vulnerable for negative stress impacts.

Although land degradation is thought to be increasing in several parts of southern Africa, the exact rate at which this is happening is uncertain. Natural drylands are the most susceptible to land degradation such as soil erosion, soil salinization, soil surface crusting and bush encroachment (Hoffman 2014). The savannahs are characterized by erratic rainfall and fragile soils, and the region supports much of southern Africa's grain and livestock production. Dryland areas cover almost all of Namibia and Botswana and an estimated 50% of South Africa, large parts of southern Angola, southern Mozambique, southern Zambia, southern Tanzania and western Zimbabwe. The annual income loss due to desertification (loss of soil productivity) is estimated at several billion dollars (US) per annum (UNEP 1999). Consumption is a major cause of land degradation (UNEP 2013) which is also a major cause of biodiversity loss.

In almost all southern African countries, policy on land could, in general, be described as piecemeal, blurred or non-existent as Mushala and Pilime stated in 1994. Since then, there have been positive developments in the sense that policies have been developed and laws accepted by parliament. In practice, however, many existing policies lack implementation, i.e. some of the countries seem not to have the resources to implement policies or may even lack the political will to do so. It should also be noted that in, e.g. South Africa, there are numerous legislative acts which deal with environmental issues. These range from the Bill of Rights to specific norms and standards for contaminated land.

In some countries like in South Africa, legal frameworks are well developed. In their review on 25 years of soil environment-oriented research, Schoeman and Van Deventer (2004) stated that environmental issues have gained importance while quoting Section 24 of the Bill of Rights in the Constitution of South Africa (Republic of South Africa 1996). "Everyone has the right: (a) to an environment that is not harmful to their health or well-being; (b) to have the environment protected, for the benefit of present and future generations, through reasonable legislative and other measures that (i) prevent pollution and ecological degradation, (ii) promote conservation, and (iii) secure ecologically sustainable development and use of natural resources while promoting justifiable economic and social development".

From a legislative viewpoint, some strides have been made. Specifically, in South Africa the National Norms and Standards for the Remediation of Contaminated Land and Soil Quality (South Africa 2014) was published with the following purposes:

(a) Provide a uniform national approach to determine the contamination status of an investigation area.

(b) Limit uncertainties about the most appropriate criteria and methods to apply in the assessment of contaminated land.
(c) Provide minimum standards for assessing necessary environmental protection measures for remediation activities.

A most relevant standard in the context of this review is the "Soil Screening Value 1" (SSV1) which means soil quality values that are protective of both human health and ecotoxicological risk for multiexposure pathways, inclusive of contaminant migration to the water resource. Also included in this policy is "Soil Screening Value 2" which is the same as SSV1 but in the absence of water. This SSV2 is applicable to three different locations, viz. SSV2 informal residential (unplanned settlement using makeshift structures on land with no infrastructure), SSV2 standard residential (formal settlement on serviced land with architectural planning) and SSV2 commercial/industrial.

There are, however, glaring differences between the lowest protective level, SSV1 vs SSV2 informal residential and SSV2 standard residential sites, as highlighted by Maboeta and Fouché (2014) looking at Cu as an example. They concluded that the relation between these benchmarks and real-world levels is at variance and more so with ecotoxicological assessments in the light of a SSV1 of 16 $\mu g\ g^{-1}$ of Cu vs 1,100 $\mu g\ g^{-1}$ for informal and 2,300 $\mu g\ g^{-1}$ for standard residential sites. It is therefore relevant to see if standardized ecotoxicological test (e.g. OECD and ISO) utilizing soil organisms can be used to assess these benchmark values in terms of ecological relevance for southern African conditions. In addition to this, it must be highlighted that no mention is made about the protection of soil or soil fauna in particular but only water, residential areas and industry. This might be an important new challenge and direction to be explored by policymakers, in South Africa in particular.

2.3 *Soil Contamination Due to Mining and Mine Waste Disposal*

Mining plays a significant role in the economies of several southern African countries where both physical and chemical environmental impacts on ecosystems are evident. Figure 2a, b shows the active mining areas and selected mining areas by commodity in South Africa to provide an idea of where mining operations are concentrated (Fig. 3). Acid and metal contamination of soil due to mining activities and industrialization have become under increasing scrutiny. A study by Ikenaka et al. (2010) indicated that heavy metal pollution in Zambia is still increasing due to human activities, especially mining. Although mining is an important source of revenue for countries like Botswana, South Africa, Zambia and Zimbabwe, it is also a major source of environmental contamination. More than 1,700 operating mines are currently registered in South Africa (Department of Mineral Resources South Africa 2018). With the increasing search for new sources of energy, interest

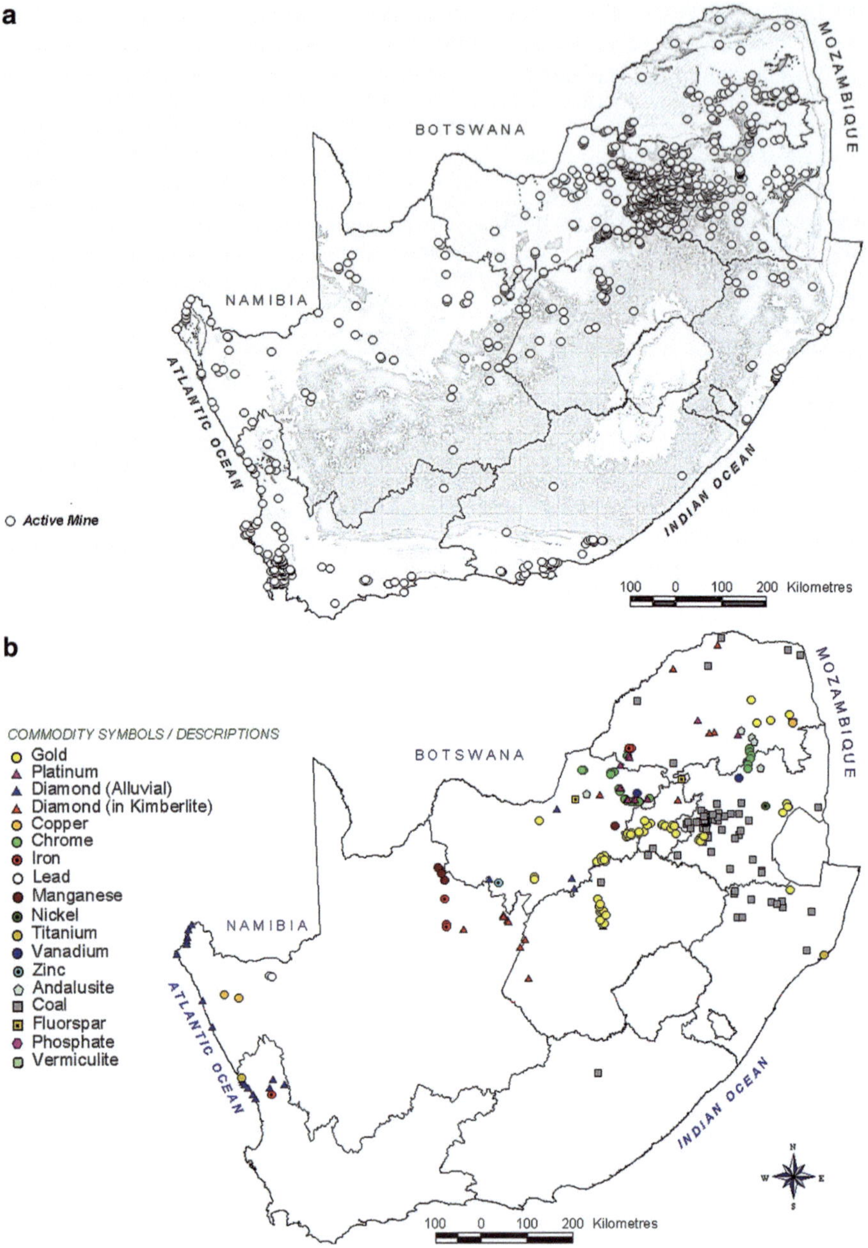

Fig. 2 (**a**) Map (adapted from source) showing active mining areas in South Africa, Lesotho and Swaziland. The colours refer to simplified geology with legend deleted since it is not important for current paper (Council for Geoscience South Africa 2018a, b). (**b**) Map (adapted from source) showing selected active mining areas per commodity in South Africa, Lesotho and Swaziland (Council for Geoscience South Africa 2018a, b)

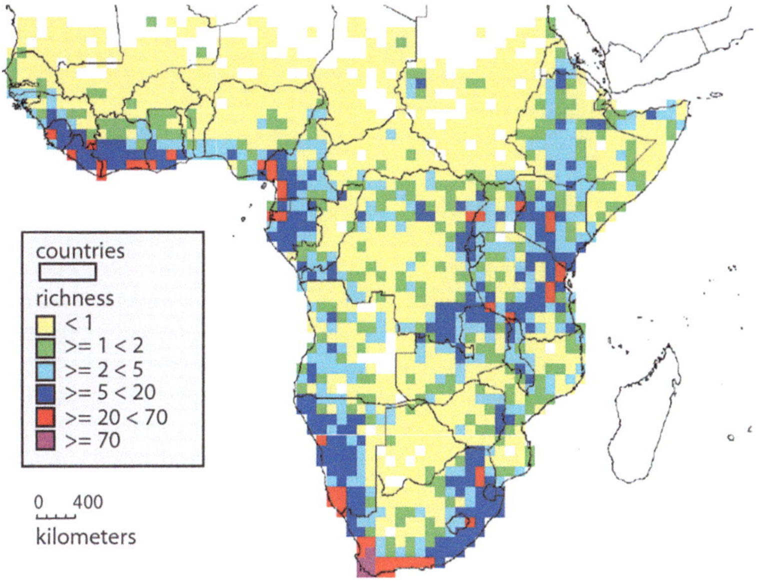

Fig. 3 Richness of endemic plant species expressed as percentage endemic species of the total number of species per 100 × 100 km block (adapted from Linder 2014) with major mining areas indicated by ovals

in the fracking process to access oil has also become a new threat to soils. Another example of a looming threat is the mining for coal in an important agricultural area in Mpumalanga (Simpson and Berchner 2017). Of the large volumes of hazardous wastes (over 418 t) estimated to have been generated only in South Africa during 1997/1998, 90% came from mining (Department of Environmental Affairs and Tourism DEAT 2006). The secondary sources of contamination from mining activities, such as soil erosion and water leaching into the surrounding areas, can largely increase the effects of mining activities on the surroundings. According to Fairbanks et al. (2000), 4.9% of South African land has been degraded with mines and quarries contributing 0.14% (175,421 ha). As only 12% of South Africa's soil is arable (Governmental Communication and Information System 2013), the extent to which mines and quarries can affect this asset (details not known) could have a huge negative impact on food security for the whole country.

A limited number of studies from southern Africa have focused on the effects of mining on soil. These include effects of, e.g. platinum mining in South Africa (Jubileus et al. 2013; Maboeta et al. 2008; Rauwane et al. 2018; Wahl et al. 2012) and tailings dams in Zimbabwe (Meck 2013) and South Africa (Blight et al. 2012).

There have been only a few studies on background or reference levels of metal contamination in the southern African countries (Herselman 2007; Herselman et al. 2005). Ekosse (2009) conducted a study in Botswana, and Ikenaka et al. (2010) carried out a similar survey for Zambia although on a smaller scale, while Ayeni et al. (2010) specifically sampled a water catchment area (river beds and accompanying riverbanks) for a regional survey around Cape Town.

The contamination of areas surrounding mining complexes and smelters was studied by Ettler et al. (2011) and Vitkova et al. (2011a, b) at the Nkana copper smelter in Zambia. Similarly did Ekosse (2005a) and Ekosse et al. (2006) around Ni-Cu smelting in Selebi-Phikwe and manganese mining in Botswana, while Rauch and Fatoki (2013) sampled the area around the platinum mines in the Bushveld Igneous Complex in South Africa. For the latter, background levels in the soil were found to be 2.7 ± 0.4 μg g^{-1} and 0.6 ± 0.2 μg g^{-1} in grass, compared to a maximum of 256 ± 122 μg g^{-1} in grass at the smelter site. Gold mining in South Africa resulted in vast volumes of tailings being deposited in impoundments, and poor management of some tailings dams resulted in seepage, adversely affecting soils and water quality (Rösner and Van Schalkwyk 2000). Catastrophic flow failures have occurred with alarming frequency in mine tailings dams and dumps but also in dumps of municipal solid waste and even in what were intended to be controlled landfills. Apart from the environmental devastation caused by these flows, they are also dangerous to human life and society (Blight and Fourie 2005). Their impact on soil biodiversity has not been studied in depth.

Apart from these mining locations, tailing disposal facilities (TDFs) derive special attention. These consist of intensively treated (physically crunched and chemically extracted) ore minerals. This results in a fine-grained slurry mixture with low pH and organic matter content which is deposited over large areas of many km^2. After draining, these PDFs consist of unstructured soil material without vegetation – revegetation may take many years – which is prone to water and wind erosion. Due to this erosion, heavy metal-loaded soil particles become dispersed over a wide area. Kaonga and Ebenso (2011) measured 180 μg m^{-3} of PM10 particles in the air close to a gold mine and 78 μg m^{-3} close to a platinum mine. PM10 particles with a size ≤ 10 μm are thought to be a major cause of respiratory problems. Aerial transport of these fine particles may occur over vast distances (kilometres). The potential adverse effects of such ultrafine particles on plants and animals have hardly been investigated so far.

2.4 Background Levels and Heavy Metal Behaviour Studies

For a proper risk assessment, it is necessary to have good insight into the natural base line levels of contaminants compared to levels at contaminated sites where observed impacts are assessed. Although for man-made chemicals, this seems to be no issue – levels should be zero – it is an issue for naturally occurring chemicals like heavy metals and for polycyclic aromatic hydrocarbons which are formed by burning of organic matter either naturally or man-induced. These kinds of background studies are limited in the western world and even less in southern Africa. Herselman et al. (2005) analysed Cd, Co, Cr, Cu, Ni, Pb and Zn contents in 4500 soil samples collected from all over the country as part of the Natural Resources Land Type mapping project. The samples were taken from the A horizon (25–40 cm thick) and represent the main soil forms described in the Land Type Memoirs (Land Type Survey Staff 1986). From these data they analysed the relation between HM content and soil factors (pH, soil organic C% and CEC). The resulting scatter diagrams were reduced to single line correlation plots and singular correlation coefficients which showed different patterns for the various HM and environmental factors – from linear and asymptotic to s-curves. This illustrates that there is not one single generic relation between soil factors and HM content. Specific knowledge, sometimes local, about the particular soil type has to be taken into consideration as was done for the soil screening values in South Africa (South Africa 2014). Ayeni et al. (2010) also studied the heavy metal contents in river beds and accompanying river banks in the surroundings of Cape Town, South Africa. The ranges that Ekosse et al. (2005), Herselman (2007) and Ayeni et al. (2010) measured are summarized in Table 1.

These values rely, however, considerably on the analytical method used and the way to calculate the final ranges; see, for instance, the difference between the total content and EDTA-extractable contents as found by Herselman (2007). Calculated ranges were done by using the quotient and product of the geometric mean and the square of the geometric standard deviation, as recommended by Chen et al. (1999, 2001), including data below the instrument detection limit as suggested by Gilbert (1987). This is important as the absolute maximum values can be considerably higher.

Herselman et al. (2006) studied sewage sludge on dedicated land disposal sites and observed some degree of leaching for Cr, Co, Ni, Cu, Zn, Cd and Pb, although this resulted in a surprisingly shallow penetration (100–200 mm) given the low pH and clay content of the studied soil materials.

Roadsides have been a topic of study in Botswana (Mmolawa et al. 2011; Shaikh et al. 2006) and in South Africa (Okonkwo and Maribe 2004) with enriched levels of Cd, Pb, Cu and Zn though with limited dispersion to some tens of metres into the kerb. Based on an inventory of roadside areas along the major highways of

Table 1 Background ranges in Botswana and South Africa (SA) for a series of heavy metals ($\mu g\ g^{-1}$) based on total or extractable analysis compared to Soil Screening Value 1 (South Africa 2014) protective of both human health and ecotoxicological risk for multiexposure pathways, inclusive of contaminant migration to the water resource (SSV1)

		Cd	Co	Cr	Cu	Ni	Zn	Pb
SSV1		7.5	300	46,000[a]	16	91	240	20
Ekosse et al. (2005)[b] Botswana	Min	0.01	1.0	0.03	11	19	17	–
	Max	0.05	28	14	116	120	68	–
Herselman (2007) SA	Min	0.62	1.51	5.82	2.98	3.43	12.0	2.59
	Max	2.7	68.5	353	117	159	115	65.8
Herselman (2007) EDTA; SA	Min	0.89	0.64	0.87	0.81	0.57	0.62	0.93
	Max	1.17	16.1	4.52	10.6	9.78	6.03	11.9
Ayeni et al. (2010) SA	Min	0.0	0.2	0.3	0.3	0.02	2.4	0.97
	Max	9.3	2.7	2.1	45.9	2.6	211.5	71.7

[a]Chromium (III)
[b]Ekosse found almost exactly the same ranges for the HM contents of the leaves of *Colophospermum mopane*

Botswana, with background levels taken 500 m from the roads, Mmolawa et al. (2011) calculated as Herselman (2007) did enrichment ratios, contamination factor, pollution load index and geo-accumulation index for a series of heavy metals (Al, Co, Cu, Fe, Mn, Ni, Pb and Zn).

In addition, the behaviour of heavy metals under variable conditions has been studied by Vitkova et al. (2011a, b) for electrostatic precipitated dust of the Mufulira copper smelter in Zambia. The release of Cd, Co, Cu, Ni, Pb and Zn was the highest at lower pH levels (3–4.5); the normal pH of the dust suspension is 4.3. As a consequence, 40% of the Cu is leached, especially when the dust contained higher proportions of fine grains. Also, Ekosse (2005a) and Ekosse et al. (2005), studying the Ni-Cu smelter at Selebi-Phikwe (Botswana), measured increased leaching levels due to the low acidity of the mine materials after mining and smelting. Due to the low CEC of the material, the ions remain in solution and thus are available for plant uptake and leaching to the groundwater.

An indirect way of measuring the available fraction of HMs in the soil is by sampling the HM content of surface water as illustrated by a study of Somerset et al. (2012). Although intended to show the efficacy of an electrochemical sensor, their studies of the aquatic environment around a platinum mining area illustrated that the available amounts of HMs in the water in general were low. Seasonal fluctuations of run-off water had an influence on the HM concentration in the waters. High concentrations of Ni, Zn and Pb were recorded which was attributed to the release from the sediment-water interface. Further, the biota samples showed higher contents of Al, Fe, Ni, Zn and Pt in August as a consequence of higher run-off during the rainy season.

2.5 *Life in Southern African Soils*

Southern Africa has a long history of published research in soil zoology which revealed a rich biodiversity. For instance, earthworm taxonomic research in southern Africa probably started with a publication (in Latin) by Kinberg (1866). Subsequent expeditions to Africa provided material that led to identifications and many descriptions of indigenous species (Michaelsen 1900, 1908, 1913a, b, c, 1933) and some morphological observations, with a comprehensive monograph by Pickford (1937) on the acanthodriline fauna. Later, renewed attention in the late 1960s was aimed at further taxonomic descriptions and limited field surveys and quantitative ecological studies (Reinecke and Ryke 1969; Reinecke and Ljungström 1969; Ljungström 1972; Reynolds and Reinecke 1976; Reinecke and Ackerman 1977; Reinecke and Visser 1980; Zicsi and Reinecke 1992; Reinecke and Alberts 1994; Haynes et al. 2003; Horn et al. 2007). These were followed by morphological and experimental studies of introduced vermicomposting species such as *Eisenia fetida*, *Perionyx excavatus* and *Eudrilus eugeniae* (e.g. Reinecke and Viljoen 1988; Reinecke et al. 1990, 1991, 1992, 1995, 2002, 2007, 2008). This provided the groundwork for later developed soil ecotoxicological tests. Several standard bioassays and some few field studies were performed on effects of heavy metals on earthworms (Maboeta et al. 2002, Van Coller-Myburgh et al. 2014) in most cases with introduced species as test models.

Conventional taxonomic knowledge of southern African indigenous earthworms was further advanced by the major contributions of Plisko (1992a, b, 1995, 1996a, b, 1999, 2001, 2002, 2003, 2004, 2006, 2010, 2012, 2013).

Apart from the contribution by Lawrence (1953) on the fauna of forest floors, the soil-dwelling *Acarina* of southern Africa only started receiving special attention in the 1950s when the late PAJ Ryke initiated soil zoological research at Potchefstroom, unravelling the taxonomic diversity of mites (Meyer and Ryke 1959a, b, 1960). This was followed by quantitative work and development of extraction methods (Aucamp and Ryke 1964; Du Toit et al. 1998; Loots and Ryke 1966, 1967; Ryke and Loots 1967; Theron et al. 1969; Van Den Berg and Ryke 1967) with only a few studies on ecological aspects. Several more descriptive contributions followed, such as those of Annecke and Moran (1982), Ueckermann and Smith Meyer (1988), Van Nieuwenhuizen et al. (1994), Olivier and Theron (1998) and Olivier (2006), attesting to the rich biodiversity. The only recent studies on mesofauna as potential bioindicators in ecotoxicology (Wahl et al. 2012; Jubileus et al. 2013; Rossouw 2005) assessed environmental disturbance at platinum mines in South Africa.

Ants and termites are to some extent soil-dwelling animals although they can spend much time above ground. They differ from typical soil-dwellers such as earthworms, nematodes, velvet worms and even collembolans and isopods, exhibiting complex social behaviour, unique feeding preferences and greater mobility. They are regarded as important moderators of soil and geomorphic processes (Whitford and Eldridge 2013). They may be functionally more important in influencing soil turnover than earthworms in many parts of southern Africa.

In Africa, termites are an important component of tropical and sub-tropical ecosystems, and some areas have high densities of termite mounds. Out of a total of 50 termite genera recorded from southern Africa, only 18 have one or more known pest species (Mitchell 2002). Some termite species, however, attack a wide range of crops with estimated losses varying considerably. Members of the fungus-growing subfamily Macrotermitinae are responsible for the majority of crop damage and most of tree mortality in forestry. *Hodotermes mossambicus* is the major role player in the destruction of rangeland in the arid and semiarid areas.

Termites also serve as a source of food for other animals such as birds, aardvark and shrews. Cooper and Skinner (1979) analysed faecal samples in South Africa which indicated that termites are the most important food item in the diet of the aardwolf. Only a few studies have so far focused on the ecology of termites to assess their role in decomposition and building of soil structure. Picker et al. (2007) studied the density of giant *Microhodotermes viator* mounds in southern Africa in relation to rainfall and vegetative productivity gradients. In a recent review, Mugerwa (2015) concluded that loss of termite predators and anthropogenic influences are the principal drivers of subterranean termites' destructive behaviour that is threatening the functioning and sustainability of selected African savannah ecosystems in Uganda. A recent study by Joseph et al. (2015) showed that termite mounds in

Zimbabwe can enhance woody plant functional diversity. In spite of such informative studies, the functional roles and effective contributions of the different termite species in soil turnover, energy flow and geomorphic processes are still far from clear. Given their wide distribution, their potential role as bioindicators as well as agents of environmental change, they require closer scrutiny. Moreover, their responses to environmental contaminants have not received much attention.

Despite the wide occurrence, high population densities and diversity in southern Africa, the population ecology and biogeography of ants have not been studied sufficiently. The spread of the Argentine ant (*Linepithema humile*) in South Africa and its behaviour, life history and annual cycle under South African conditions were well-documented (Skaife 1955). More recently Mothapo and Wossler (2011, 2014) provided evidence for multiple colonization of the Argentine ant, *Linepithema humile*, in the Western Cape. The preliminary picture according to Traveset and Richardson (2006) is that invasive species can cause profound disruptions in plant reproductive mutualisms if they outcompete native ant pollinator species. Ant diversity, coexistence and myrmecophyte interactions in Namibia were documented by Campbell (2013) and indicated how much more scope there is for research in this environment. However, ants and termites have not yet featured in soil ecotoxicological studies in southern Africa despite the fact that they are numerous and often the target of chemical control measures.

Free-living and parasitic nematodes are widely distributed in southern African soils, and the parasitic forms are considered a serious pest in agricultural soils (Heyns 1971; McDonald et al. 2005; Du Preez et al. 2015). Although chemical control is normally advocated, this has the potential to cause collateral damage to many beneficial organisms. An extensive survey of maize fields in South Africa by De Waele and Jordaan (1988) showed that *P. zeae*, *P. brachyurus* and *P. minor* are common parasitic nematodes associated with maize and they have a wide occurrence. Due to a lack of experimental field studies, there are still large gaps in our knowledge of the ecology and ecotoxicology of these small soil-dwelling creatures. In particular, the free-living forms have not received as much attention as their parasitic relatives.

The use of isopods as indicators of soil systems has been recommended (Gerlach et al. 2013). Limited studies followed like the work of Barnard (1960) on soil-dwelling isopods, and little progress has so far been made to understand their diversity and ecology in the drier climates of this region. Dangerfield and Telford (1994, 1995) studied woodlice populations from Botswana, but the gaps must be filled if we want to use isopods as ecological, environmental and biodiversity indicators.

Onychophorans or velvet worms are represented by several species in southern African forest floors (Ruhberg and Hamer 2005). Daniels and Ruhberg (2010) studied the molecular and morphological variation in a South African velvet worm *Peripatopsis moseleyi* (Onychophora, Peripatopsidae) and provided evidence for cryptic speciation. Daniels and Van Wyk (2011) studied the genetic variation of the critically endangered velvet worm species *Opisthopatus roseus*, endemic to the Ngele mistbelt forest in the KwaZulu-Natal province of South Africa. Weldon et al. (2013) measured metabolic and water loss rates of two cryptic species of the genus *Opisthopatus*. Knowledge of their response to adverse environmental conditions and contamination is still lacking.

Hamer (1997) contributed to an improved understanding of millipede diversity in southern Africa, and Hamer and Slotow (2002) evaluated existing data for South African millipedes in terms of usefulness for biodiversity conservation planning and management. They revealed large gaps in the collecting coverage. A comprehensive evaluation of the ecology and conservation status of these rather unique worms in Africa is needed. At present, no information is available on how they respond to soil contamination.

Our survey revealed further a scarcity of information on the diversity and role of soil microorganisms in southern African soils, although it is generally accepted that they play a vital role in soil processes and soil fertility.

A more recent, comprehensive overview of the species diversity of soil biota in South Africa (Janion-Scheepers et al. 2016) provides evidence of the country's rich soil biodiversity, which may be representative of the region as a whole. From this extensive survey, covering all relevant species groups in soil, it becomes clear that South Africa is a megadiverse country, according to the authors, and contains a multitude of endemic species. Expanding this observation to all southern African states will reinforce this statement. Table 2 gives the numbers of species in South Africa and the percentages of species that are endemic for a selected number of species groups that are important in soil systems. A number of these species groups are used for ecotoxicological testing procedures. Based on Løkke and van Gestel (1998), it is indicated whether these representatives are used (*) or regularly used (***) for ecotoxicological testing purposes.

It is clear that South Africa alone contains far more species which are endemic than species that are similar to the species found in Europe and the United States where standardized ecotoxicological testing procedure (and species) have been developed and are used. Therefore, more research is needed to verify whether the currently applied standard test species are really representatives for all parts of the world, including southern Africa. A case in point is the use of earthworms in toxicity testing and risk assessment. In soil ecological risk assessment, they play a dominant role as test species, but the question is whether the Palearctic and Oriental species that are mostly used for testing are representative for southern African conditions and species. Southern Africa has more than 200 known indigenous earthworm species, belonging mostly to the microchaetids and acanthodrilids and not to the lumbricids which normally serve as test species. Introduced European and Oriental earthworm species are few in diversity, but they flourish in disturbed soil where horticulture and agriculture are practised. Species of the genera *Aporrectodea*, *Eisenia* and *Bimastos* (from the northern hemisphere) and *Amynthas* and *Pheretima* (from the far east) have followed the European and Asian settlers wherever they set foot in southern Africa (Ljungström 1972).

2.6 Plant-Soil Life Interaction

Although this review primarily aims at the belowground ecosystems, it cannot neglect the interactions between soil life and plants since a greater part of soil life

Table 2 Numbers of species per species group found in South Africa and percentage of endemic species compared to globally known numbers of species (Janion-Scheepers et al. 2016)

Species group	# species	% endemic
Protozoa	651	Unknown
Nematoda*	441(plant)	44
	13 (EPN)	73
Enchytraeidae**	2	50
Oligochaeta**	323	84
Tardigrada	37	89
Gastropoda	525	80
Isopoda**	127	95
Amphipoda	7	70
Diplopoda*	154	80
Chilopoda*	111	Unknown
Pauropoda	4	Unknown
Symphyla	4	Unknown
Aranea	2170[a]	60
Opiliones	208	92
Scorpoiones	101	57
Pseudoscorpiones	135	70
Oribatida**	434	ca.75
Mesostigmata	281	ca.50
Trombidiformes*	700	ca.50
Microcoryphia	22	95
Diplura	40	Unknown
Protura	2	Unknown
Collembola**	124	60
Insecta		
Blattaria	230	Unknown
Termitoidea	126	33
Embioptera	37	Unknown
Orthoptera	ca.765	Unknown
Psocoptera	ca.80	Unknown
Hemiptera	ca.270	Unknown
Neuroptera	ca.252	Unknown
Coleoptera*	ca.4000	Unknown
Scarabaeniea	491	ca.37
Diptera	ca.1553	Unknown
Lepidoptera	ca.6800	Unknown
Formicidae	541	33

The species used (*) and regularly used (**) for ecotoxicological testing are marked with asterisks
[a]1331 species related to soil

is dependent on plants. This relation can be direct such as plant-parasitic nematodes, symbiotic living vesicular-arbuscular mycorrhiza (VAM or AM) or rhizosphere microorganisms. It can also be indirect in the way plant roots influence the soil substratum or the composition of dead plant material which is a main food source for detritivorous soil organisms. Soil organisms can also play a decisive role in dispersal and survival of plant species. The important role of ants in pollination and seed dispersal, especially in the fynbos of the Western Cape and also in Zimbabwe (Linzey and Washok 2000), has been recognized. Myrmecochory is common to the fire-prone shrublands of the south-western Cape. Elaiosome-bearing seeds are located rapidly by ants and transported to nests where they are protected from granivorous rodents, desiccation and fire (Witt and Giliomee 2004). Because of these direct and indirect relationships, we therefore also give a summarized overview of the research on heavy metal impacts on plant species in southern African countries.

In line with the observations above on representativeness of standard test species for the assessment of impacts on southern African plant life, plant life in southern Africa, similar to soil life, shows an enormous biodiversity. Moreover, a considerable part is endemic with the fynbos as the most prominent example. According to Linder (2001), 70% of the fynbos species are endemic. In general, the whole coastal zone of southern Africa from the south-western coastal part of Angola, the coastal zone of Namibia and South Africa to the south-eastern coast of Mozambique is characterized by high levels of endemicity (20–70%, according to Fig. 3 in Linder 2014). Further, it should be mentioned that Madagascar, also part of southern African countries, is characterized by a high percentage of endemic plant species, although ecotoxicological studies from this country are not available.

Our inventory on plant-oriented ecotoxicological research (see Sect. 3.4.2) reveals that most of the plant species studied, in general, have a broad distribution which includes contaminated areas. Quite a number of studies deal with plant species that are used as traditional medicine, while a few studies deal with plants utilized as food or fodder. Only a few papers concentrate on hyperaccumulation characteristics (Mesjasz-Przybylowicz et al. 2004), but information of its prevalence is lacking (Turnau and Mesjasz-Przybylowicz 2003).

3 Studies on Effects of Heavy Metals

3.1 Using Biomarkers

Biomarker tests are used to detect changes on the sub-organism level in living organisms, such as on the molecular, cellular and tissue levels. These changes usually occur long before changes on whole organism levels can be observed (Maboeta et al. 2002, 2003). Detrimental effects exerted on organisms by pollutants can be identified shortly after exposure has occurred by using these methods which act as early warning tools (Peakall 1992). Biomarkers have already been used as diagnostic tools in the medical field for a long time, and since the late 1970s, soil scientists have started implementing these tests as indicators of toxicological exposure in animals (Stenersen 1979, 1980; Depledge and Fossi 1994; Dallinger 1996; Svendson et al. 1996; Svendson and Weeks 1997a, b; Morgan et al. 1999; Arnaud et al. 2000). Since the turn of the millennium, biomarker research has proliferated, and since 1999, soil scientists in southern Africa have started using biomarkers to identify and measure responses of organisms to exposures such as pesticides, heavy metals and contaminated soils (Reinecke and Reinecke 1999a, b; Reinecke and Reinecke 2007a, b; Snyman et al. 2000; Maboeta et al. 2002, 2018).

South African ecotoxicologists have been involved in many projects where biomarkers were used to investigate toxic effects on soil organisms. Apart from research done in southern Africa by collaborating southern African scientists, several studies involved international collaborators (Reinecke et al. 2002: Aamodt et al. 2007; Maleri et al. 2008b; Van Gestel et al. 2009).

Biomarkers have been used in tests with different soil organisms such as earthworms, isopods and snails to assess damage caused by various heavy metals. Responses were studied using biomarkers such as the neutral red retention test (NRRT) measuring a lysosomal response in cells (Reinecke et al. 2002; Reinecke and Reinecke 2003; Maboeta et al. 2002, 2003, 2004), the neutral red spectrophotometric assay (NRR) (Maleri et al. 2008a), the MTT spectrophotometric test (Voua Otomo and Reinecke 2010), the comet assay (Reinecke and Reinecke 2004; Fourie et al. 2007; Voua Otomo et al. 2014; Fouché et al. 2016), testing cell viability with Trypan Blue (Snyman and Odendaal 2009) as well as the evaluation of tissue damage caused by toxicants (Reinecke et al. 1995; Odendaal and Reinecke 2007).

The NRRT assay measured in an indigenous species *Microcheatus* sp., exposed to the copper contained in the fungicide copper oxychloride in the field, showed increased cellular damage with increased exposure concentrations. This also correlated with a decrease in population density with increased toxicant concentrations (Maboeta et al. 2002). Similar results were obtained using the earthworm species *Aporrectodea caliginosa* exposed to the same copper-containing pesticide in field plots (Maboeta et al. 2003). The effect of copper oxychloride was also studied in the laboratory on the ubiquitous earthworm species, *Eisenia fetida/andrei* (Maboeta et al. 2004; Reinecke et al. 2002). The toxic impact on soils receiving effluents from the gold, chrome, copper, manganese, platinum and vanadium mines is a concern (Jubileus et al. 2013; Van Coller-Myburgh et al. 2014, 2015). The NRRT biomarker was used by the latter authors to study these effects, using *E. andrei* as the animal model. It was a sensitive marker of damage caused by the metals and other substances contained in the mine tailings. In a study by Maboeta and Fouché (2014), NRRT was used as an endpoint to determine effects of copper pollution due to copper-smelting activities on *E. andrei*. It was also a very sensitive endpoint which revealed good indications of effects of copper over a gradient of soil stretching from the smelter to distances of 5 km away from it.

Effects of different concentrations of lead on *E. fetida*, pre-exposed during culturing to lead, were also studied in the laboratory with the NRRT technique. Reinecke and Reinecke (2003) found that lead had a detrimental effect on the life cycle of the worms. Similar studies have been carried out on this earthworm species exposed for a number of years to Zn and Cd heavy metals (Reinecke and Reinecke 1999b). Clear differences in NRRT results between control and metals exposed worms were shown, indicating that this biomarker can be used effectively to show historic exposure to these metals in worm populations. Using the same principle of the leaking time of neutral red from lysosomes (NRRT), a spectrophotometric assay (NRR) was done for the evaluation of cytotoxicity in *E. andrei*. Worms were exposed to a gradient of Cd concentrations in ultramafic soils (soils with naturally high concentrations of some metals). Varying responses were obtained using this fast spectrophotometric NRR method (Maleri et al. 2008a). The effect of Cd was also tested on earthworms using another spectrophotometric assay, the MTT test. It is based on the principle of the reduction of an MTT substrate (yellow) to a blue formazan salt by metabolically active cells. This biomarker was used in a study by Voua Otomo and Reinecke (2010) indicating an increased cytotoxic tolerance to Cd after a long-term exposure of *E. andrei* to the metal.

The Trypan Blue exclusion assay to determine cell viability was used by Snyman and Odendaal (2009) who utilized this technique successfully as a biomarker of cellular damage in the woodlouse *Porcellio laevis* exposed to cadmium. The results showed statistically significant differences in percentage viable cells and exposure concentrations in the animals.

Biomarkers measured on the molecular level have also been used in southern Africa. The comet or single cell gel electrophoresis assay measures DNA damage in the form of DNA strand breaks in the cells of organisms exposed to toxic substances. Such damage can compromise the genetic integrity of affected organisms. Reinecke and Reinecke (2004) assessed the genotoxic risk of nickel pollution on the terrestrial environment in South Africa using the earthworm species *E. fetida*. Their results showed that this heavy metal had genotoxic potential with a clear dose relationship of low to high damage with increasing Ni concentrations.

Fourie et al. (2007) used this genotoxicity biomarker to determine species sensitivities of a number of earthworm species to cadmium. They used two exotic species routinely used in ecotoxicity tests, *E. fetida* and *A. caliginosa*, as well as some exotics not often cultured in South Africa for vermicomposting, i.e. *Amynthas diffringens* and *Dendrodrilus rubidus* and an indigenous species *Microchaetus benhami*. The comet biomarker proved to be very effective to demonstrate DNA damage in three of the five species tested. The study showed clear differences in species' sensitivities to cadmium which could serve as another application of this biomarker test. Although these findings should be followed up, they seem to indicate that there are clear differences between the responses of the indigenous and exotic species. This again poses the question whether the "standardized" lumbricid species commonly used in toxicity testing and risk assessment can serve as a representative of the local southern African soil fauna.

In the paper by Voua Otomo and Reinecke (2010), the comet assay was used as a second biomarker on tolerance of *E. andrei* to cadmium. Voua Otomo et al. (2014) used the comet assay to determine whether Cd and Zn had genotoxic effects on the earthworm *E. fetida* and whether these will differ from being exposed to a mixture of the two substances. In this experiment, the earthworms were also exposed to different temperatures in the soil, and the effect of this added factor on genotoxicity was measured. It was found that the genotoxicity of Cd increased with increasing temperature bringing a new dimension to the impacts on soil animals in the light of predicted future temperature changes as a result of climate change. Reinecke and Reinecke (1997) showed that the heavy metals, lead and manganese caused ultrastructural damage to the sperm of *E. fetida*. Odendaal and Reinecke (2007) also showed damage to the hepatopancreas of the isopod *Porcellio laevis* on exposure to zinc with degrees of damage quantitatively revealed.

It can be concluded that, although the correlation between biomarker results and higher-level effects of contaminants is still not well established for practical application in regulatory assessments, the results obtained thus far are promising (Maboeta et al. 2002, 2003, 2004). The possibility of using these results on biomarker responses of soil organisms to predict effects and changes at higher levels of organization – whole organism and population levels – has been discussed extensively in a review paper by Reinecke et al. (2007).

3.2 Studies on Soil Microorganisms

The various articles dealing with microbial effects can be sorted into three groups: one related to plant-fungi studies mainly on VAM fungi (also called AM), another group on ecotoxicological impacts on microbial numbers and enzyme activities and a third group on microbial breakdown studies. The last group only dealt with microbial degradation of cyanide (Motaung et al. 2012) and poly aromatic hydrocarbons (PAHs) by Atagana (2004).

Turnau et al. (2013) studied the internal distribution of Ni, Cr, Co, Fe and Ti in the fern plants *Pellaea viridis* growing on ultramafic rock. With regard to plant-fungi studies, in plant parts hosted by the fungal endophyte *Glomus tenue*, increased amounts of these metals were measured, while the plant avoided uptake. This suggests that fungi might play a role in immobilization of metals in plant tissues. Orlowska et al. (2011) discussed mycorrhizal colonization as related to plant establishment (the shrub *Dodonaea viscosa* and the grasses *Andropogon eucomus* and *Imperata cylindrica*) on an alkaline gold mine tailing in South Africa. Mycorrhizal infection significantly increased the survival of the plants. Kanu et al. (2013) investigated the easy infection of *Aspalathus linearis* (rooibos) with AM fungi and hypothesizes on the further possibilities for phyto-accumulation, phytostabilization and phyto-degradation, although without further specification or quantitative underpinning. Spruyt et al. (2014) assessed the status of trees used for phytoremediation of mine waste sites (Au, Ur, Zn and Pt mining). *Tamatrix* sp., *Searsia lancea* and *Searsia pendulina* planted on waste sites were investigated for the presence of AM fungi by performing molecular RNA analysis which confirmed moderate to high colonization rates. This may assist in future initiatives when AM fungi are being used as remediation agents. Straker et al. (2007) also looked at the AM status of soils and tailings around gold and uranium mines in relation to *Asclepias fruticosa*, *Cynodon dactylon*, *Atriplex semibaccata*, *Phytolacca octandra* and *Asparagus laricinus*. *Asclepias fruticosa* and *Asparagus laricinus* showed the highest AM status on sludge deposits (also called slime dams) irrespective of their revegetation status, whereas *Cynodon dactylon* and *Atriplex semibaccata* performed the best in the "veld" (surrounding area).

Du Plessis et al. (2005) undertook a laboratory study of the response of microbes (protistan numbers) to increasing concentrations of copper oxychloride, regularly used in viticulture. Different growth media were selectively used for heterotrophic bacteria, actinomycetes, pseudomonads and fluorescent bacteria. A negative impact was already observed with the addition of 15 mg kg^{-1} EDTA-extractable Cu. Claassens et al. (2006) studied the microbial structure of coal mine waste. A positive relation was found between microbial biomass, vegetation cover, and nutrient contents of the soil cover. Moreover, the various sites of different ages were characterized by a different microbial composition: the youngest sites had the highest abundances of fungi and actinomycetes. Maboeta et al. (2006) investigated soil from a platinum mine which revealed an increasing level of beta-glucosidase, urease, phosphatase and dehydrogenase enzyme activity when moving further away from the mine site. Van Coller-Myburgh et al. 2014 found significant negative impacts of chrome waste on dehydrogenase, beta-glucosidase, acidic and alkaline phosphatase and urease in the laboratory, which may impact C, P and N cycles in the soil. Due to the complex composition of the mine waste, it was not possible to relate these negative impacts to a specific chemical element in the waste.

A somewhat different approach was followed by Kouadjo and Zeze (2011) in their study of the Cr tolerance of *Staphylococcus* species in a fly ash dump site where they found 67 chromium-resistant bacteria, some of which could reduce chromium. Botes et al. (2007) did a similar study on As resistance by bacteria from an antimony mine and found two species capable of reducing arsenate to arsenite, especially *Serratia marcescens* which was able to sustain extremely high As contents in liquid cultures (up to 500 mmol L^{-1}), although the addition of arsenate or arsenite resulted in the temporarily reduction in the growth rate and extended lag phases of the test species.

In conclusion, there is quite some research about the role of VA fungi in the accumulation of HMs by specific plant species. Consequences for phytoremediation are hypothesized though not worked out quantitatively. To what extent this increased HM load might negatively influence consumers of plant parts – either natural plant feeders or human consumers of food or their use for medicinal purposes – is hardly known. Knowledge about the impact of HMs on microbial species' structure or enzyme activities is limited but confirms the negative impact of HMs. Given the limited number of papers about HM impacts, it is not possible to assess the consequences of these impacts for the major ecological element cycles of C, N and P in the soil.

3.3 Studies on Soil Fauna

It has already come to light from previous studies that limited information is available on the current impact of mining on the soil mesofauna (Eijsackers et al. 2014, 2017). This statement is also true for all studies pertaining to soil invertebrate ecotoxicological studies, especially as far as heavy metals, the focus of this overview, are concerned.

The organisms mostly focused on are those proposed by international guidelines which include earthworms (Van Coller-Myburgh et al. 2014, 2015; Otomo et al. 2016; Fouché et al. 2016; Maboeta and Fouché 2014), enchytraeids (Voua Otomo et al. 2013a, b; Otomo et al., 2014), snails (Snyman and Odendaal 2009; Snyman et al. 2000, 2002) and woodlice (Odendaal and Reinecke 1998, 1999a, b, c, 2003, 2004a, b, c) with only a limited few focusing on soil mesofauna (Jubileus et al. 2013; Wahl et al. 2012).

With regard to earthworm studies, many of these focused on metal-based fungicides (Eijsackers et al. 2005; Maboeta et al. 2003, 2004; Reinecke et al. 2002; Vermeulen et al. 2001; Helling et al. 2000), metal salts (Voua Otomo and Reinecke 2010; Owojori et al. 2009a, b, c; Fourie et al. 2007; Reinecke et al. 1997, 2001, 2002; Maboeta et al. 1999; Reinecke and Reinecke 1997) and mining (Van Coller-Myburgh et al. 2015, Van Coller-Myburgh et al. 2014; Maboeta and Fouché 2014; Jubileus et al. 2013; Maboeta et al. 2008). The overriding conclusions of these studies were that growth and reproduction of earthworms proved to be a valuable tool in in the assessment of metal-based fungicides, metal salt and mine tailings containing a cocktail of different metals. These included the standardized species *Eisenia fetida/andrei* as well as exotics (e.g. *Aporrectodea caliginosa*) and endemics (*Microcheatus* sp.). The studies on *E. fetida/andrei* were all laboratory-based studies, while the other were all field based. Both Helling et al. (2000) and Reinecke et al. (2002) found that copper oxychloride had negative effects on life cycle parameters of *E. fetida* as was also corroborated by Maboeta et al. (2003, 2004). The findings of these laboratory-based studies were followed up with field trials and including exotics (e.g. *Aporrectodea caliginosa*) and endemics (*Microcheatus* sp.) during which similar results were observed in terms of biomass and numbers of worms in experimentally sprayed plots and vineyards (Eijsackers et al. 2005).

Voua Otomo et al. (2013a, b) and Voua Otomo and Reinecke (2010) reported on the use of enchytraeids in the assessment of metal-spiked soil as well as soils collected from gold mines. It was found that the enchytraeid reproduction test could be a useful tool for the relatively rapid assessment of metal-polluted soils. This is especially suited for assessing metal contaminated sites in emerging economies. This suit of tests can become a valuable complimentary addition to the well-established and researched earthworm tests which are mostly used in bioassays of polluted soils. There is, however, a need for comparative studies between earthworms and enchytraeids to investigate whether they should both be used to complement each other, or alternatively, to utilize enchytraeids as an initial test to determine potential risks and following it up with an earthworm test.

Odendaal and Reinecke (2004) undertook experimental studies on the effects of toxic metals on woodlice from areas close to smelters and found them to be suitably sensitive experimental animals for ecotoxicological studies. There is a need to determine the distribution, population densities and biodiversity of isopods in southern African soils where metal-containing biocides are used since field studies are lacking. These gaps must be filled if we want to use isopods as ecological, environmental and biodiversity indicators.

In conclusion, compared to the western world, there are not extensive contributions on the ecotoxicological use of soil invertebrates in southern Africa. The available studies deal with a limited number of investigated and tested species carried out by only a few researchers. Most of these studies focused on the effects of contaminants on agricultural soils, mining soils and other industries.

The scope of the research ranged from effects at the molecular level and cellular biomarkers to individual/population dose effects. No studies have investigated effects at the ecosystem level with only a few contributions on field earthworm populations (Maboeta et al. 2002, 2003) and soil mesofaunal diversity (Rossouw 2005; Jubileus 2008; Wahl et al. 2012). Most of these investigations were based on regulatory guidelines (OECD and ISO) with few on naturally occurring species in southern African habitats.

3.4 Plants and Heavy Metals

Most studies on ecological impacts of soil contaminants in southern African countries found in our selection have been carried out on plants. Most of these papers reported on the uptake and accumulation of contaminants, with a few examples of hyperaccumulation, which were aimed at remediation or restoration of contaminated areas with plants. Ecological impacts have been studied to a lesser extent, with only a few publications on interactions with fungi and also some in relation to edibility and medicinal aspects of the investigated plant species.

3.4.1 Ecotoxicological Impacts on Vegetation

Only a small number of studies dealt with the vegetation of mining areas. Mapaure et al. (2011) studied the impact of a Cu-mine dump on semiarid savannah in Kombat, Namibia. They compared the dump area containing elevated levels of As, Cr, Cu, Pb and Zn, with a control area, and observed a significantly reduced woody vegetation both with respect to species richness (18.00 ± 1.12 versus 11.83 ± 1.45) and diversity (Shannon-Wiener 2.87 ± 0.06 versus 2.36 ± 0.15). *Fingerhuthia africana* and *Solanum nigrum* were only found at the control site and were apparently sensitive, whereas *Euclea divinorum*, *Eragrostis superba*, *Aristida congesta* and *Commiphora glandulosa* were only found on the contaminated site and were apparently tolerant to the elevated HM levels. The observed differences could only be attributed to the HM contents, although other factors are also involved.

Weiersbye et al. (2006) performed an extensive survey of the vegetation of 56 gold and uranium tailing dams in the surroundings of Klerksdorp, Carletonville and Welkom in South Africa (SA). The differences between the three studied regions were considerable. Only 60 species occurred in all three regions, of which 24 species had been introduced during rehabilitation activities.

3.4.2 Monitoring Accumulation of Contaminants in Plants

Monitoring studies in the sense of repeatedly carried out inventories according to a fixed spatial and temporal pattern have not been carried out in southern Africa yet, although there have been a few one-off inventories. Odiyo et al. (2005) studied the metal contamination in Thoyoyandou, South Africa, in roadside soils and vegetation. They found a linear correlation between the metal concentrations of soil and vegetation (except for Cu) in the order of Pb > Zn > Cr > Cu > Cd for soil and Pb > Zn > Cd > Cr > Cu for vegetation. Mills et al. (2012) studied roadside trees with respect to their nutrient status but concentrated on their aerosol capture in a further tree less savannah landscape. They observed that the soil next to tree trunks had higher concentrations of Ca, Zn and Sr and to a lesser extent of As, Se and Mo compared to soil at some distance from trees. On the other hand, they showed for Be, V, Fe, Co, Ti, Cr, Mn and Ni relative impoverishment. Most of these metals indicate atmospheric accretion. Mills et al. (2012) did not comment on potential contamination risks of these processes. Okedeyi et al. (2014) inventoried the vicinity of a coal power plant, especially with respect to transfer to *Digitaria eriantha*, and found low transfer (accumulation factor < 1) for Fe, Cu, Mn, Ni, Cd, Pb, Hg, Cr and Zn, except Pb with an accumulation factor of 20. A somewhat different study, because it dealt with radionuclides which are also mined in southern Africa, was undertaken by Oyedele and Shimboyo (2013) on the concentrations of 238U, 232Th and 40K around four major towns in southern Namib, Namibia. The concentrations of 232Th and 40K were above world average levels though below permissible radiation limits for human health of these radionuclides. A number of inventories have been carried out around mines. Masvodza et al. (2013) studied a transect near a slime dam of a Zimbabwe gold mine. They analysed Ni, Cu, Fe, Zn, Pb and As contents in soil and roots, leaves and bark of *A. saligna* and *A. polyacantha* and found bioaccumulation (BAFs > 1) for Ni, Cu and Fe. In a study around an abandoned manganese mine, Ekosse et al. (2006) found only weak and negative correlations between Mn contents of soil and vegetation (leaves). Ekosse (2005a, b) also sampled the surroundings of a Ni/Cu mine in Zimbabwe for Cu, Ni, Fe, Cr, Co, Cd, Zn and Se in soil and *Colophospermum mopane* leaves. Concentrations in soil and leaves decreased with distance from the mine and smelter/concentration plant.

Boyd et al. (2008) studied the hyperaccumulation of Ni by *Senecio coronatus* and especially concentrated on the derived consequences: to what extent could hyperaccumulation influence the decomposition of *Senecio* leaves and as such influence ecosystem functions like litter breakdown in general. Nickel hyperaccumulation was found to decrease the decomposition of the leaves and increased effect in soils with elevated Ni content than control soils. Moreover, the Ni was quickly released from the leaf tissue, after 1 month 57–68% of the biomass was lost and only 9–28% of the original Ni content remained.

3.4.3 Studies on Restoration, Revegetation and Phytostabilization

The retrieved reported studies on revegetation and phytoremediation concentrated on the principles of application of plants to restore mining areas. Among other areas covered were the potential of specific plant species in combination with microorganisms and the possibilities of revegetation. A complicating factor is that in these papers, and in these types of papers in general, various terms were used with different meanings and not always with a proper definition. In order to explain the differences between the different treatment approaches, we have typified the various approaches in the following short-hand definitions:

- Restoration is the general process of bringing a degraded system as closely back to its original status.
- Revegetation is the process of appearance of different plant species in spatial coherence on bare soil material (either by natural colonization or by direct planting or sowing).
- Phytoremediation is the process by which plants remediate (clean up) a contaminated area by taking up contaminants and binding or degrading them.
- Phytostabilization is the process by which bare soil, sensitive to erosion, is bound by activities of roots and associated soil organisms and, hence, become more stable.
- Phytovolatilization is the process by which elements like Hg, Se and As are directly transpiring/evaporating into the atmosphere or transformed to volatile, harmless compounds above as well as underground.

An overview of these different processes and relevant plant species was given by Koptsik (2014), including a quantitative assessment related to the situation in practice and what plants can achieve with respect to remediation and stabilization. The phytoremediation of HM-contaminated soil by specific hyperaccumulators not only depends on the percentage of HM taken up but by the actual amounts of HM in the biomass of the plant that can be harvested and removed from the contaminated area compared to the total amount of HM left in the soil. Liphadzi and Kirkham (2005) described phytoremediation as a technology, but concentrated on the general chemical behaviour of HMs and the plant physiological processes involved and did not specify any southern African plant species or location. Remediation processes on the potential of *Aspalathus linearis* (rooibos) as an indicator of the remediation of Al contaminated soils were described by Kanu et al. (2013). The article also briefly discussed the environmental and economic implications of remediation. Masvodza et al. (2013) studied the accumulation of Ni, Cu and Fe from a gold mine sludge dam in Zimbabwe by *Acacia saligna* and *A. polyacantha*. From the BAF (bioaccumulation factor), they concluded that these species could potentially be used for remediation. The potential for remediation is partly related to the presence and role of mycorrhizal fungi, as mentioned by Kanu et al. (2013) for rooibos.

Most papers on revegetation described this process as such. Dafana et al. (2010) studied the early growth performance of *Acacia* species (*A. gerrardii*, *A. karroo* and *A. polyacantha*) on nickel mine tailings in Zimbabwe. In a randomized block design, they compared these three species and three treatments (top soil, sewage sludge and compound fertilizer) but did not find any significant difference between species and treatments and concluded that all species can be used for revegetation. Blignaut and Milton (2005) studied the impact of clumping on survival of three succulent plant species (*Aridaria noctiflora* spp. *noctiflora*, *Drosanthemum deciduum* and *Psilocaulon dinteri*) on gypsiferous mine waste. From their findings, 1 year after replanting on the mine waste, 67% of the plants had survived, though the plants planted individually grew better than the three species planted together, presumably as a result of competition for resources. Morgenthal et al. (2004) studied the recolonization of chrysotile tailings by recording crown cover of native grassland species adapted to these circumstances for 8 years. Application of natural topsoil improved the recolonization. After initial establishment of *Melinis repens*, it was foreseen, given the natural successional sequence, that the vegetation could develop to grassland with dominating perennial grasses (*Hyperthelia dissoluta*, *Cymbopogon excavatus*, *Themeda triandra* and *Heteropogon contortus*). Mentis (2006) assessed the processes for revegetation using native grassland of former open cast coal-mining areas and concentrated on the depletion of soil organic carbon and nitrogen and how to improve these conditions.

An inventorying approach was used by Weiersbye et al. (2006) in their study of the floristic composition of a series of gold and uranium mine tailings in South Africa. From their survey of 56 tailings comprising 5,864 ha and a similar area with tailings-polluted soils, they recorded 375 plant species. Earlier studies resulted in 86 additional species. The most common species groups were Poaceae (107 species), Asteraceae (81 species), Fabaceae (55 species) and Anacardiaceae (16 species). Of these species 76% was indigenous, although there were clear differences between the three regions surveyed; only 60 species occurred in all 3 regions, of which 24 species had been introduced for rehabilitation purposes. From an ecological perspective, 53–88% of the species, varying according to the type of substrate, could be characterized as persisters or natural colonizers. The majority of these were indigenous and perennial (76% and 85%) and large shrubs or trees. Annuals were only observed on recently vegetated sites. Most species were perennial, and annual species were rare on old vegetated sites. Of all the species, only 4% (mostly alien herbaceous species) had also been introduced by rehabilitation activities. As the main reasons for the presence of all species, the authors mention their tolerance of acid mine drainage and salinity. Weiersbye et al. (2006) commented that there were many indigenous plant species suitable for restoration of mine tailings given their tolerance, though unfortunately the authors did not provide data on which plant species would have been present naturally in these areas so that possible negative impacts could be assessed. Westcott (2011) studied the germination of three grass species (*Anthephora pubescens*, *Cynodon dactylon* and *Panicum maximum*) on clayey and sandy soils as well as on platinum and gold mine tailings. In general, the germination was found to be low, also when the seeds were treated with a coating stimulating germination and establishment. In the field experiments, they did even better in tailings, but mainly due to the absence of competition from other establishing plant species.

3.4.4 Metals in Edible and Medicinal Plants

Several studies have been conducted on the uptake of HMs by food crops, especially the ones cultured in home or school gardens. Although this has no direct relation with ecotoxicological impacts, the similarity in general performance between, for instance, cultured and wild Cruciferae could be used for a first assessment of ecological risks. Bvenura and Afolayan (2012) measured Cu, Mn, Zn, Pb and Cd in cabbage, carrot, onion, spinach and tomato from a home garden in the Eastern Cape. The study revealed very high levels of Mn and Zn levels exceeding safe values for human consumption set by the national health authorities. In addition, Kootbodien et al. (2012) sampled school gardens near mine tailing dams in Johannesburg and found that the estimated daily intake of Pb, Hg and As for a child of 30 kg was within the acceptable health limits. Mahlangeni et al. (2012) studied the metal uptake in sweet potatoes and found that at some sites in the vicinity of Durban, levels of As and Pb exceeded permissible levels set for food consumption. Furthermore, Lion and Olowoyo (2013) studied uptake for *Spinacia oleracea* experimentally in waste dump soil from Tshwane (SA) and found considerably elevated Cu and Zn concentrations with Hazard Quotients above permissible levels for human consumption. A few studies dealt with medicinal plants or plants used as a nutritional supplement like *Moringa oleifera* investigated by Pakade et al. (2013). These farm-cultivated plants contained no toxic metals; only Ca and Mg showed increased concentrations. Quite a number of studies indicated that the plant species investigated were used for medicinal treatment. The consequences of the consumption of these plants within general elevated HM contents on human health were hardly discussed however. A study by Street et al. (2009) on the effect of cadmium uptake and accumulation on the growth and antibacterial activity of *Merwilla plumbea* showed that low levels of Cd (2 mg L^{-1}) not only reduced growth of leaves, bulbs and roots but also resulted in increased Cd contents of roots and bulbs. In addition, bulb extracts showed reduced antibacterial activity. This accumulation as well as the reduced antibacterial activity might influence the medicinal value of the bulbs.

We conclude that there are only very few studies on the adverse impacts of HMs on natural plant species and revegetation. These show, however, clearly that the HM amounts on and around mining areas have a considerable negative impact on the indigenous flora. On the other hand, there were quite some examples of restoration of mine deposits by planting or sowing of resistant (nonindigenous) plant species. The results of the investigations with VAM indicated that the adverse influence of HMs on VAM also hampered establishment of plant species on mine deposits. With respect to edible plants, most studies did not show adverse effects, while HM impacts on medicinal effects of plant species have not been reported.

4 Conclusions

At a country scale, looking at the extent and geographical distribution of ecotoxicological research emanating from southern Africa, the research activity is very unevenly distributed. The numbers of publications on heavy metal research that we found for the different countries were relatively low compared to similar studies for Europe or the United States. In southern Africa, it was highest for countries with stronger economies and/or developed mining industries such as South Africa. The relatively low research activity might not be a concern for countries which currently have no known major problems with heavy metal contamination. However, experience elsewhere has shown that population growth and industrialization may change the environmental impacts in these areas and this could require more active research in the field of ecotoxicology.

Increasing urbanization, more traffic and especially mining are the main sources for the increasing heavy metal contamination in the environment. For the countries with a considerable mining industry, with large cities and the related traffic congestions, the scientific publications recorded and found during our literature surveys were Zimbabwe 20, Zambia 17, Botswana 17 and Namibia 4 papers. No relevant research papers were found for the Democratic Republic of Congo. The amount of reported research, as reflected by the numbers of papers, is limited considering the potential environmental and health risks posed by heavy metals in these areas. Thus, for a major part of southern Africa, we have to rely on research often undertaken elsewhere under vastly different conditions to assess the related risks (Eijsackers et al. 2017).

As an overlying process, we have to realize that climate change will not only have a direct negative impact due to changing environmental conditions (Erasmus et al. 2002) but also some indirect impacts. Due to these changed conditions, the availability of contaminants in the soil might change, and consequently their impacts on soil life could be enhanced.

In order to address the current situation, a better ERA (ecological risk assessment) procedure should be developed for southern African conditions. The procedures used at present are not sufficiently attuned to the specific biota and environmental conditions in southern Africa as we have shown previously (Eijsackers et al. 2017). However, in countries with such overwhelming large and diverse natural areas, abundant wildlife and varying landscapes, it would not be sufficient just to use a preventive ERA procedure aimed at individual chemical compounds. An assessment is needed of the total toxic load emanating from different sources and its impact on the very diverse ecosystems in southern Africa.

Zooming in, as a next step, we have to compare the regional distribution of contamination with the biodiversity of the contaminated and potentially contaminated areas. To illustrate this, we have used the map of Linder (2014) for endemic plant species richness (Fig. 3). This map can be compared to Fig. 2a, b where mining areas are indicated to give an idea on the impact of mining on these areas of species richness. According to Linder's research, the biodiversity distribution (total species richness) for plant species greatly coincides with those of mammals, birds, snakes and frogs. We further assume that accordingly plant and soil life will have roughly the same distribution in biodiversity richness because of the expected aboveground-belowground associations (Wardle et al. 2004). When we compare this distribution with the distribution for mining activities, as given for South Africa, it is clear that most mining generally occurs in areas with a relatively low biodiversity. However, it is not ecological sound reasoning to conclude that species in these areas are thus not important. They may still play important functional roles in these particular ecosystems, especially because these areas have highly selective conditions, resulting in a species diversity of partly relatively resistant and partly highly adapted species. These species, due to these adaptations, have a specialized position in the total species diversity and, moreover, are less adapted to other adverse conditions and as such deserve special protection. In general, it can be assumed that ecosystems with a low biodiversity could therefore be more susceptible to adverse impacts by contamination (e.g. by mining activities). Especially along the coastal lines, a high to very high plant species diversity is observed, characterized by high numbers of endemic plant species (Linder 2001).

Hence, both in areas with high biodiversity and with high endemism, as regularly occurs, in southern Africa (Linder 2001 for plant species and Janion-Scheepers et al. 2016 for soil animals), and areas with low biodiversity, it is necessary to protect sensitive species and execute proper risk assessment. This will require assessment procedures applicable to these local conditions and faunal and floral composition (Eijsackers et al. 2017).

It is therefore important to analyse and assess whether there is sufficient knowledge of these environmental conditions and biodiversity to carry out such risk assessments.

In general, there is a fair picture of the HM contamination, especially at point locations, but not for more diffuse contamination. With respect to background levels, there is some information, but nothing about gradients and about specific, soil-related, high levels. Given the results of Herselman (2007), there are no generalized relations between HM content and environmental factors. This makes it problematic and indeed complicated to execute a proper ecological risk assessment for different, specific locations.

Much is known about impacts of heavy metals on plants, and the picture does not differ so much between the outcomes of research in the various southern African countries.

Vertebrates were not included in this review although the number of publications dealing with metals and vertebrates was surprisingly low. The studies on wild grazers that we found do indicate, however, that the impacts could be considerable and deserve further attention.

Information about food chain accumulation of metals in the region is also very limited, although the general literature on heavy metal impacts shows that it can be expected to be considerable. Similarly, there is hardly any information on the influence of HMs on ecosystem relations. As an illustration, no research reports were found dealing with the role of metals in plant-insect associations, which are very important in ecosystem functioning.

More specifically with relation to soil, the main focus of this review, we can conclude that:

- There is a fair body of knowledge about the southern African soil micro-, meso- and macrofauna and microorganisms, although most of it relates to taxonomic descriptions and laboratory experiments, while field studies and validations are scarce. Further, the number of researchers dealing with soil organisms, especially in relation to environmental contamination, is extremely small.
- There is limited information about microorganisms both with respect to structural (total and species numbers of bacteria and fungi) and functional aspects (enzyme levels but also new promising research fields like metabolomics). Also, with respect to VAM, most important in relation to availability of nutrients and contaminants for plants, the knowledge is limited.
- There is quite a body of knowledge at the biochemical, physiological and genetic level in relation to soil fauna which can be usefully applied in biomarker studies.

5 Summary and Future Outlook

In spite of all this knowledge, there is clearly a need to integrate the available information on soil life into an integrated and workable soil ecosystem assessment procedure that could be used in southern Africa. It is difficult to assess the ecotoxicological risks for the soil ecosystem properly because of our given limited knowledge of the many interrelations between organism groups and the variability of environmental conditions. This confirms the need for more soil ecological studies.

It is possible to derive generalized hazards from the international literature, but extrapolating and tuning these hazards to more specific regional risk conditions would be complicated for the following reasons:

- Southern Africa is characterized by different and often more extreme environmental conditions than those in Europe or the United States where the generalized hazards have been derived; many species are endemic, and we do not know the extent of their sensitivity or tolerance ranges for heavy metals. More specific research on these aspects is needed.

As further activities, we therefore suggest the following:

1. The performance of an ERA for specific natural areas, tailor made and attuned to the local conditions and fauna (Eijsackers et al. 2014). Development of at least a guideline on how to carry out such attuned ERAs. This could include an assessment of the (possibly higher) sensitivity of endemic species in contaminated soils like in mining areas.
2. As an illustration of such an ERA directed at natural areas and landscapes, Linder's classification (Linder 2014) for a number of vegetation zones could be used to establish correlations with faunal characteristics. Similar studies at a more appropriate finer-scaled regional level like the chapter by Jonas et al. (2006) on mining and other impacts in Mucina and Rutherford's book on vegetation of South Africa, Swaziland and Lesotho (map on p. 746) could also be used.
3. As a next step, an inventory could be executed of the contaminated areas (e.g. mining) that are within natural and conservation areas and consecutively followed up with the performance of ERAs for the prominent food chains and target species in those areas, in combination with an ERA for general ecological functions. This will need further studies, but the elements for such an ERA are for the greater part available.
4. This blueprint of an ERA has to be mirrored with the instruments contained in legislation directed at the proper management of areas surroundings mining areas. So far the direct surroundings of mines and technical buildings are covered by the already-established SSV1 limits, but this has to be extended to include the wider surroundings that are affected by wind-driven dust. Wind erosion is also a serious problem for waste disposal sites, some of which are located in the direct vicinity of urban and agricultural areas. Hence, more soil conservation measures such as phytostabilization are needed. This approach could provide for the proper implementation of an ecologically sound and sustainable policy for the management of southern African soils and its rich biodiversity.

Acknowledgements This work was funded by the National Research Foundation (NRF) of South Africa and the Unit of Environmental Sciences of the North-West University. Opinions expressed and conclusions arrived at are those of the authors alone. The authors would also like to thank Oluwatosin Oladipo.

References

Aamodt S, Kontestabo HS, Sverdrup LE, Gubrandsen N, Reinecke SA, Reinecke AJ, Stenersen J (2007) Recovery of cholinesterase activity in the earthworm *Eisenia fetida* Savigny following exposure to chlorpyrifos. Environ Toxicol Chem 26:1963–1967

Annecke DP, Moran VC (1982) Insects and mites of cultivated plants in South Africa. Butterworths, Durban

Arnaud C, Saint-Denis M, Narbonne JF, Solar P, Ribera D (2000) Influence of different standardised test methods on biochemical responses in the earthworm *Eisenia fetida andrei*. Soil Biol Biochem 32:76–73

Atagana HI (2004) Co-composting of PAH-contaminated soil with poultry manure. Lett Appl Microbiol 39(2):163–168

Aucamp JL, Ryke PAJ (1964) A preliminary report on a grease film extraction method for soil microarthropods. Pedobiologia 4:77–79

Ayeni OO, Ndakidemi PA, Snyman RG, Odendaal JP (2010) Metal contamination of soils collected from four different sites along the lower Diep River, Cape Town, South Africa. Int J Phys Sci 5(12):2045–2051

Barnard KH (1960) Terrestrial Isopoda from the Transvaal. Ann Natal Mus 15(2):45–55

Blight GE, Fourie AB (2005) Catastrophe revisited–disastrous flow failures of mine and municipal solid waste. Geotech Geol Eng 23(3):219–248

Blight G, Copeland A, Jardine P, MacRobert C (2012) Measurements on freshly-deposited surfaces of two platinum tailings dams. J South Afr Inst Min Metall 112(11):911–917

Blignaut A, Milton SJ (2005) Effects of multispecies clumping on survival of three succulent plant species translocated onto mine spoil in the Succulent Karoo Desert, South Africa. Restor Ecol 13(1):15–19. https://doi.org/10.1111/j.1526-100X.2005.00003.x

Botes E, Heerden EV et al (2007) Hyper-resistance to arsenic in bacteria isolated from an antimony mine in South Africa. S Afr J Sci 103(7/8):279–281

Boyd RS, Davis MA et al (2008) Does hyperaccumulated nickel affect leaf decomposition? A field test using Senecio coronatus (Asteraceae) in South Africa. Chemoecology 18(1):1–9. https://doi.org/10.1007/s00049-007-0387-7

Bvenura C, Afolayan AJ (2012) Heavy metal contamination of vegetables cultivated in home gardens in the Eastern Cape. S Afr J Sci 108(9–10):1–6

Campbell H (2013) Ant diversity, coexistence and myrmecophyte interactions in Namibia. PhD diss., University of Reading

Chen M, Ma LQ, Harris WG (1999) Baseline concentrations of 15 trace elements in Florida surface soils. J Environ Qual 28:1173–1181

Chen M, Ma LQ, Hoogeweg CG, Harris WG (2001) Arsenic background concentrations in Florida, U.S.A. surface soils: determination and interpretation. Environ Forensics 2:117–126

Claassens S, Rensburg PJJV et al (2006) Soil microbial community structure of coal mine discard under rehabilitation. Water Air Soil Pollut 174(1/4):355–366. https://doi.org/10.1007/s11270-006-9125-y

Cooper RI, Skinner JD (1979) Importance of termites in the diet of the aardwolf *Proteles cristatus* in South Africa. S Afr J Zool 14:5–8

Council for Geoscience South Africa (CGSA) (2018a) Active mines South Africa, Lesotho and Swaziland. http://www.geoscience.org.za/images/Maps/activemines.gif. Accessed 21 Nov 2018

Council for Geoscience South Africa (CGSA) (2018b) Selected active mines South Africa, Lesotho and Swaziland. http://www.geoscience.org.za/images/Maps/selectedactivemines.gif. Accessed 21 Nov 2018

Dafana M, Nyakudya IW et al (2010) Comparative analysis of the early growth performance of indigenous Acacia species in revegetating Trojan nickel mine tailings in Zimbabwe. Elec J Env Agricult Food Chem Title 9(8):1393–1403

Dallinger R (1996) Metallothionein research in terrestrial invertebrates: synopsis and perspectives. Comp Biochem Physiol 113(C):125–133

Dalvie MA, Africa A, London L (2009) Change in the quantity and acute toxicity of pesticides sold in South African crop sectors, 1994–1999. Environ Int 35(4):683–687

Dangerfield JM, Telford SR (1994) Population size structure and sex ratios in some woodlice (Crustacea: Oniscidae) from southern Africa. J Trop Ecol 10:261–271

Dangerfield JM, Telford SR (1995) Tactics of reproduction and reproductive allocation in four species of woodlice from southern Africa. J Trop Ecol 11:641–649

Daniels SR, Ruhberg H (2010) Molecular and morphological variation in a South African velvet worm Peripatopsis moseleyi (Onychophora, Peripatopsidae): evidence for cryptic speciation. J Zool 282(3):171–179

Daniels SR, van Wyk JH (2011) Genetic variation in the critically endangered velvet worm Opisthopatus roseus (Onychophora: Peripatopsidae). Afr Zool 46(2):419–424

De Waele D, Jordaan EM (1988) Plant-parasitic nematodes on field crops in South Africa. 1. Maize. Rev Nématol 11:65–74

Department of Mineral Resources South Africa. http://www.dmr.gov.za/. Accessed 20 June 2018

Department of Environmental Affairs and Tourism (DEAT) (2006) A report on the state of the environment in South Africa. Department of Environmental Affairs and Tourism, Pretoria

Depledge MH, Fossi MC (1994) The role of biomarkers in environmental assessment (2). Invertebrates. Ecotoxicology 3:161–172

Du Plessis KR, Botha A et al (2005) Response of the microbial community to copper oxychloride in acidic sandy loam soil. J Appl Microbiol 98(4):901–909. https://doi.org/10.1111/j.1365-2672.2004.02537.x

Du Preez G, Swart A, Fourie H (2015) Nematodes of the Wonderfontein Cave (Witwatersrand Basin, South Africa). Nematology 17(8):967–980

Du Toit BJ, Theron PD, Ueckermann EA (1998) A new genus and four new species of the family Camerobiidae (Acari: Raphignathoidea) from South Africa. Int J Acarol 24(1):3–19

Eijsackers H, Beneke P, Maboeta M, Louw JPE, Reinecke AJ (2005) The implications of copper fungicide usage in vineyards for earthworm activity and resulting sustainable soil quality. Ecotoxicol Environ Saf 62(1):99–111

Eijsackers H, Maboeta M, Doelman P (2006) Vital soil as basis for sustainable soil management, a survey of soil problems in The Netherlands and the way to tackle it, with an assessment for the South African situation. Suid-Afrik Tydskr Nat Wet Tegnol 25(4):300–317

Eijsackers H, Reinecke A, Reinecke S, Maboeta M (2017) Threatened southern African soils: a need for appropriate ecotoxicological risk assessment. Environ Impact Assess Rev 63:128–135

Eijsackers H, Swartjes FA, Van Rensburg L, Maboeta MS (2014) The need for attuned soil quality risk assessment for non-Western humans and ecosystems, exemplified by mining areas in South Africa. Environ Sci Policy 44:174–180

Ekosse G (2005a) Heavy metals concentrations in the biophysical environment around the Ni-Cu mine and the smelter/concentrator plant, Selebi Phikwe, Botswana. Glob J Environ Sci 4(2):97–110

Ekosse GI (2005b) Petrographic and physico-chemical characterisation of tailings dump and soils around a nickel-copper mining and smelting environment. Glob J Environ Sci 4(1):31–39

Ekosse GIE (2009) Multivariate analyses and spatial distribution of manganese minerals in soils close to an abandoned manganese mine. Land Degrad Dev 20(3):283–299. https://doi.org/10.1002/ldr.912

Ekosse EGI, Ngila JC et al (2005) Multivariate analyses of heavy metals in soils and Colophospermum mopane leaves around the Selebi Phikwe nickel-copper mine and smelter/concentrator plant area, Botswana. J Appl Sci Environ Manag 9(1):177–185

Ekosse GIE, Fouche PS et al (2006) Total organic carbon in soils and its relation with manganese concentrations in soils and vegetation close to an abandoned manganese mine. Int J Environ Sci Technol 3(1):15–24

Erasmus BF, Van Jaarsveld AS, Chown SL, Kshatriya M, Wessels KJ (2002) Vulnerability of South African animal taxa to climate change. Glob Chang Biol 8(7):679–693

Ettler V, Mihaljevic M et al (2011) Tracing the spatial distribution and mobility of metal/metalloid contaminants in Oxisols in the vicinity of the Nkana copper smelter, Copperbelt province, Zambia. Geoderma 164(1/2):73–84. https://doi.org/10.1016/j.geoderma.2011.05.014

Fairbanks D, Thompson M, Vink D, Newby T, Van den Berg H, Everard D (2000) South African land-cover characteristics database: a synopsis of the landscape. S Afr J Sci 96:69–82

Fey M (2010) Soils of South Africa. Cambride University Press, Cambridge

Fouché T, Maboeta M, Claassens S (2016) Effect of biofumigants on soil microbial organisms and ecotoxicology of earthworms (*Eisenia andrei*). Water Air Soil Pollut 6:227–256

Fourie F, Reinecke SA, Reinecke AJ (2007) The determination of earthworm species sensitivity differences to cadmium genotoxicity using the comet assay. Ecotoxicol Environ Saf 67:361–368

Gerlach J, Samways M, Pryke J (2013) Terrestrial invertebrates as bioindicators: an overview of available taxonomic groups. J Insect Conserv 17:831–850

Gilbert RO (1987) Statistical methods for environmental pollution monitoring. Van Nostrand Reinhold, New York

Herselman JE, Steyn CE, Fey MV (2005) Baseline concentration of Cd, Co, Cr, Cu, Pb, Ni and Zn in surface soils of South Africa: research in action. S Afr J Sci 101(11–12):509–512

Governmental Communication and Information System (2013) SA yearbook 2012/13 agriculture, forestry and fisheries. GCIS, Pretoria

Hamer ML (1997) A preliminary assessment of the southern African millipede fauna: diversity and conservation (Diplopoda). Entomol Scand Suppl 51:209–221

Hamer ML, Slotow RH (2002) Conservation application of existing data for South African millipedes (Diplopoda). Afr Entomol 10(1):29

Haynes RJ, Dominy CS, Graham MH (2003) Effect of agricultural land use on soil organic matter status and the composition of earthworm communities in KwaZulu-Natal, South Africa. Agric Ecosyst Environ 95(2):453–464

Helling B, Reinecke SA, Reinecke AJ (2000) Effects of the fungicide copper oxychloride on the growth and reproduction of Eisenia fetida (Oligochaeta). Ecotoxicol Environ Saf 46(1):108–116

Herselman JE (2007) The concentration of selected trace metals in South African soils. PhD thesis, University of Stellenbosch, Stellenbosch

Herselman JE, Steyn CE et al (2006) Dedicated land disposal of wastewater sludge in South Africa: leaching of trace elements and nutrients. Water Sci Technol 54(5):139–146. https://doi.org/10.2166/wst.2006.556

Heyns J (1971) A guide to the plant and soil nematodes of South Africa. Balkema, Cape Town

Hoffman MT (2014) Changing patterns of rural land use and land cover in South Africa and their implications for land reform. J South Afr Stud 40(4):707–725

Horn JL, Danuta Plisko J, Hamer ML, Griffiths CL (2007) The leaf-litter earthworm fauna (Annelida: Oligochaeta) of forests in Limpopo Province, South Africa: diversity, communities and conservation. Afr Zool 42(2):172–179

Ikenaka Y, Nakayama SMM et al (2010) Heavy metal contamination of soil and sediment in Zambia. Afr J Environ Sci Technol 4(11):729–739

Janion-Scheepers C, Meassey J, Brachler B, Chown SL, Wilson JRU (2016) Soil biota in a megadiverse country: current knowledge and future research directions in South Africa. Pedobiologia 99:129–174

Jonas Z, Rouget M, Reyers B, Mohamed B, Rutherford MC, Mucina L, Powrie LW (2006) Vulnerability assessment of vegetation types. In: Mucina L, Rutherford MC (eds) The vegetation of South Africa, Lesotho and Swaziland, Strelitzia, vol 19. South African National Biodiversity Institute, Pretoria

Joseph G, Makumbe M, Seymour CL, Cumming GS, Mahlangu Z, Cumming DHM (2015) Termite mounds mitigate against 50 years of herbivore-induced reduction of functional diversity of savanna woody plants. Landsc Ecol 30:2161–2174

Jubileus MT (2008) Assessment of platinum mine tailings storage facilities: an ecotoxicological perspective. MSc dissertation, North-West University, Potchefstroom Campus

Jubileus MT, Theron PD, Van Rensburg L, Maboeta MS (2013) Utilising *Eisenia andrei* to assess the ecotoxicity of platinum mine tailings disposal facilities. Ecotoxicology 22:331–338

Kanu SA, Okonkwo JO et al (2013) Aspalathus linearis (Rooibos tea) as potential phytoremediation agent: a review on tolerance mechanisms for aluminum uptake. Environ Rev 21(2):85–92. https://doi.org/10.1139/er-2012-0055

Kaonga B, Ebenso E (2011) An evaluation of atmospheric aerosols in Kanana, Klerksdorp gold mining town, in the North-West Province of South Africa. In: Mazzeo N (ed) Air quality monitoring, Assessment and management. InTech, Rijeka. http://www.intechopen.com/books/air-quality-monitoring-assessment-and-management/an-evaluation-of-atmospheric-aerosols-in-kanana-klerksdorp-gold-mining-town-in-the-north-west-provin

Kinberg JGH (1866) Annulata nova. Öfversigt af Förhandlingar Kongelige Vetenskaps-Akademiens 22:239–258
Kootbodien T, Mathee A et al (2012) Heavy metal contamination in a school vegetable garden in Johannesburg. S Afr Med J 102(4):226–227
Koptsik GN (2014) Problems and prospects concerning the phytoremediation of heavy metal polluted soils: a review. Eurasian Soil Sci 47(9):923–939. https://doi.org/10.1134/S1064229314090075
Kouadjo CG, Zeze A (2011) Chromium tolerance and reduction potential of Staphylococci species isolated from a fly ash dumping site in South Africa. Afr J Biotechnol 10(69):15587–15594. https://doi.org/10.5897/AJB10.078
Land Type Survey Staff (1986) Land types of the map 2730 Vryheid. In: Memoirs on the agricultural resources of South Africa no. 7. ARC–Institute for Soil, Climate and Water, Pretoria
Lawrence RF (1953) The biology of the cryptic fauna of forests. With special reference to the indigenous forests of South Africa. Balkema, Cape Town
Linder HP (2001) Plant diversity and endemism in sub-Saharan tropical Africa. J Biogeogr 28(2):169–182
Linder HP (2014) The evolution of African plant diversity. Front Ecol Evol 2:1–14. https://doi.org/10.3389/fevo.2014.00038
Linzey AV, Washok KA (2000) Seed removal by ants, birds and rodents in a woodland savanna habitat in Zimbabwe. Afr Zool 35(1):295–299
Lion GN, Olowoyo JO (2013) Population health risk due to dietary intake of toxic heavy metals from Spinacia oleracea harvested from soils collected in and around Tshwane, South Africa. S Afr J Bot 88:178–182
Liphadzi MS, Kirkham MB (2005) Phytoremediation of soil contaminated with heavy metals: a technology for rehabilitation of the environment. S Afr J Bot 71(1):24–37
Ljungström PO (1972) Introduced earthworms of South Africa. On their taxonomy, distribution, history of introduction and on the extermination of endemic earthworms. Zool Jahr Abt Syst 99:1–81
Løkke H, van Gestel CA (1998) Handbook of soil invertebrate toxicity tests. Wiley, Chichester
Loots GC, Ryke PAJ (1966) A comparative, quantitative study of the micro-arthropods in different types of pasture soil. Zool Afr 2(2):167–192
Loots GC, Ryke PAJ (1967) Ratio of Oribatei-Trombidiformes with reference to organic matter content in soils. Pedobiologia 7(2–3):121–124
Maboeta MS, Reinecke AJ, Reinecke SA (1999) The effects of low lead levels on the growth and reproduction of the African earthworm *Eudrilus eugeniae* (Oligochaeta). Biol Fertil Soils 30(1–2):113–116
Maboeta MS, Fouché T (2014) Utilising an earthworm bioassay (*Eisenia andrei*) to asses a South African soil screening value with regards to effects from a copper manufacturing industry. Bull Environ Contam Toxicol 93:322–326
Maboeta MS, Reinecke SA, Reinecke AJ (2002) The relation between lysosomal biomarker and population response in a field population of *Microchaetus sp.* (Oligochaeta) exposed to the fungicide copper oxychloride. Ecotoxicol Environ Saf 52:280–287
Maboeta MS, Reinecke SA, Reinecke AJ (2003) Linking lysosomal biomarker and population responses in a field population of *Aporrectodea caliginosa* (Oligochaeta) exposed to the fungicide copper oxychloride. Ecotoxicol Environ Saf 56:411–418
Maboeta MS, Reinecke SA, Reinecke AJ (2004) The relationship between lysosomal biomarker and organismal responses in an acute toxicity test with *Eisenia fetida* (Oligochaeta) exposed to the pesticide copper oxychloride. Environ Res 96:95–101
Maboeta MS, Van Rensburg L, Jansen van Rensburg PJ (2008) Earthworm (*Eisenia fetida*) bioassay to assess the possible effects of platinum tailings disposal facilities on the environment along a gradient. Appl Ecol Environ Res 6(2):13–19

Maboeta MS, Claassens S et al (2006) The effects of platinum mining on the environment from a soil microbial perspective. Water Air Soil Pollut 175(1/4):149–161. https://doi.org/10.1007/s11270-006-9122-1

Maboeta MS, Oladipo OG, Botha SM (2018) Ecotoxicity of mine tailings: unrehabilitated versus rehabilitated. Bull Environ Contam Toxicol 100(5):702–707

Mahlangeni N, Moodley R et al (2012) Soil nutrient content on elemental uptake and distribution in sweet potatoes. Int J Veg Sci 18(3):245–259. https://doi.org/10.1080/19315260.2011.628369

Maleri RA, Fourie F, Reinecke AJ, Reinecke SA (2008a) Photometric application of the MTT- and NRR-assays as biomarkers for the evaluation of cytotoxicity *ex vivo* in *Eisenia andrei*. Soil Biol Biochem 40:1040–1048

Maleri RA, Reinecke AJ, Reinecke SA (2008b) Metal uptake of two ecophysiologically different earthworms (Eisenia fetida and Aporrectodea caliginosa) exposed to ultramafic soils. Appl Soil Ecol 38(1):42–50

Mapaure I, Chimwamurombe PM, Mapani BS, Kamona FA (2011) Impacts of mine dump pollution on plant species diversity, composition and structure of a semiarid savanna in Namibia. Afr J Range Forage Sci 28(3):149–154

Masvodza DR, Dzomba P, Mhandu F, Masamha B (2013) Heavy metal content in *Acacia saligna* and *Acacia polyacantha* on Slime Dams: implications for phytoremediation. Am J Exp Agric 3 (4):871–883

McDonald AH, Loots GC, Fourie H, De Waele D (2005) A microplot study on Ditylenchus africanus population densities and damage symptoms on groundnut in relation to commercial yields. Nematology 7(5):647–653

Meck ML (2013) Geochemistry for sustainable development in Africa: Zimbabwe case study. In: Chemistry for sustainable development in Africa. Springer, Berlin, Heidelberg, pp 105–122

Mentis MT (2006) Restoring native grassland on land disturbed by coal mining on the Eastern Highveld of South Africa. S Afr J Sci 102(5/6):193–197

Mesjasz-Przybylowicz J, Migula P et al (2004) Ecophysiology of Chrysolina pardalina Fabricius (Chrysomelidae), a herbivore of the south African Ni hyperaccumulator Berkheya coddii (Asteraceae). In: Boyd RS, Baker AJM, Proctor J (eds) Ultramafic rocks: their soils, vegetation and fauna. Science Reviews 2000 Ltd, St Albans. Proceedings of the Fourth International Conference on Serpentine Ecology, Cuba, 21–26 April, 2003; 2004 233–241 33 ref

Meyer MK, Ryke PAJ (1959a) Mites of the superfamily Raphignathoidea (Acarina: Prostigmata) associated with South African plants. J Nat Hist 2:209–234

Meyer MKPS, Ryke PAJ (1959b) A revision of the spider mites (Acarina: Tetranychidae) of South Africa, with descriptions of a new genus and new species. J Entomol Soc South Afr 22(2):330–366

Meyer MKP, Ryke PAJ (1960) Mites of the superfamily Eupodoidea (Acarina: Prostigmata) associated with South African plants. S Afr J Agric Sci 3:481–496

Michaelsen W (1900) Oligochaeta. In: Das Tierreich. Lief. 10. R. Friedländer & Sohn, Berlin, pp 1–575

Michaelsen W (1908) III. Annelida. A. Oligochäten aus dem Westlichen Kapland. In: Schultze L (ed) Zoologische und Anthropologische Ergebnisse Forschungsreise im westlichen und zentralen Südafrika. Bd 1. Lief. 2, Denkschriften der medizinisch-naturwissenschaftlichen Gesellschaft zu Jena, vol 13, pp 30–42

Michaelsen W (1913a) Die Oligochaeten des Kaplandes. Zool Jahr Abt Syst 34:473–556

Michaelsen W (1913b) The Oligochaeta of Natal and Zululand. Ann Natal Mus 2(4):397–457

Michaelsen W (1913c) Oligochaeten vom tropischen und südlich-subtropischen Afrika. Teil II. Zoologica 68:1–63

Michaelsen W (1933) Opisthophore Oligochäten aus dem mittleren und dem südlichen Africa hauptsächliche gesammelt von Dr F Haas während der Schomburgh-Expedition 1931–1932. Abh Senckenberg Naturforsch Ges 40(4):409–433

Mills AJ, Milewski AV et al (2012) Aerosol capture by small trees in savannahs marginal to treeless grassland in South Africa. Geoderma 189(190):124–132. https://doi.org/10.1016/j.geoderma.2012.05.006
Mitchell JD (2002) Termites as pests of crops, forestry, rangeland and structures in Southern Africa and their control. Sociobiology 40:47–69
Mmolawa KB, Likuku AS, Gaboutloeloe GK (2011) Assessment of heavy metal pollution in soils along major roadside areas in Botswana. Afr J Environ Sci Technol 5(3):186–196
Morgan AJ, Stürzenbaum SR, Kille p. (1999) A short overview of biomarker strategies with particular regard to recent developments in earthworms. Pedobiologia 43:574–584
Morgenthal T, Maboeta M et al (2004) Revegetation of heavy metal contaminated mine dumps using locally serpentine-adapted grassland species. (Special issue: Stress physiology). S Afr J Bot 70(5):784–789
Motaung TE, Albertyn J, Kock JL, Pohl CH (2012) *Cryptococcus cyanovorans* sp. nov., a basidiomycetous yeast isolated from cyanide-contaminated soil. Int J Syst Evol Microbiol 62 (5):1208–1214
Mothapo NP, Wossler TC (2011) Behavioural and chemical evidence for multiple colonisation of the Argentine ant, Linepithema humile, in the Western Cape, South Africa. BMC Ecol 11(1):6
Mothapo NP, Wossler TC (2014) Resource competition assays between the African big-headed ant, Pheidole megacephala (Fabricius) and the invasive Argentine ant, Linepithema humile (Mayr): mechanisms of inter-specific displacement. Ecol Entomol 39(4):501–510
Mugerwa S (2015) Magnitude of the termite problem and its potential anthropogenic causes in Nakasongola district of Uganda. Grassl Sci 61(2):75–82
Mushala HM, Pilime AT (1994) Guidelines for land evaluation and land use planning in the SADC region. Unpublished consultancy report for the Southern African Development Community (SADC) -ELMS and SADC-L&WMRP Sections, June 1994
Odendaal JP, Reinecke AJ (1998) The effect of high lead concentrations on the mortality, mass and behaviour of *Porcellio laevis* Latr. (Crustacea, Isopoda) in laboratory tests. Afr Zool 33(3):143–146
Odendaal JP, Reinecke AJ (1999a) The sublethal effects and accumulation of cadmium in the terrestrial isopod *Porcellio laevis* Latr. (Crustacea, Isopoda). Arch Environ Contam Toxicol 36 (1):64–69
Odendaal JP, Reinecke AJ (1999b) The toxicity of sublethal lead concentrations for the woodlouse, *Porcellio laevis* (Crustacea, Isopoda). Biol Fertil Soils 29(2):146–151
Odendaal JP, Reinecke AJ (1999c) Short-term toxicological effects of cadmium on the woodlouse, *Porcellio laevis* (Crustacea, lsopoda). Ecotoxicol Environ Saf 43(1):30–34
Odendaal JP, Reinecke AJ (2003) Quantifying histopathological alterations in the hepatopancreas of the woodlouse *Porcellio laevis* (Isopoda) as a biomarker of cadmium exposure. Ecotoxicol Environ Saf 56(2):319–325
Odendaal JP, Reinecke AJ (2004a) Effect of metal mixtures (Cd and Zn) on body weight in terrestrial isopods. Arch Environ Contam Toxicol 46(3):377–384
Odendaal JP, Reinecke AJ (2004b) Evidence of metal interaction in the bioaccumulation of cadmium and zinc in *Porcellio laevis* (Isopoda) after exposure to individual and mixed metals. Water Air Soil Pollut 156(1):145–161
Odendaal JP, Reinecke AJ (2004c) Bioaccumulation of cadmium and zinc, and field validation of a histological biomarker in terrestrial isopods. Bull Environ Contam Toxicol 72(4):769–776
Odendaal JP, Reinecke AJ (2004) Bioaccumulation of cadmium and zinc, and field validation of a histological biomarker in terrestrial isopods. Bull Environ Contam Toxicol 72(4):769–776
Odendaal JP, Reinecke AJ (2007) Quantitative assessment of effects of zinc on the histological structure of the hepatopancreas of terrestrial isopods. Environ Contam Toxicol 53:359–364
Odiyo JO, Bapela HM et al (2005) Metals in environmental media: a study of trace and platinum group metals in Thohoyandou, South Africa. Water SA 31(4):581–588

Okedeyi OO, Dube S et al (2014) Assessing the enrichment of heavy metals in surface soil and plant (Digitaria eriantha) around coal-fired power plants in South Africa. Environ Sci Pollut Res 21(6):4686–4696. https://doi.org/10.1007/s11356-013-2432-0

Okonkwo JO, Maribe F (2004) Assessment of lead exposure in Thohoyandou, South Africa. Environmentalist 24(3):171–178

Olivier PAS (2006) A first record of the family Penthalodidae Thor, 1932 (Acari: Prostigmata) from South African soils, with descriptions of two new species in the genus Stereotydeus Berlese, 1901. Afr Entomol 14(1):53

Olivier PAS, Theron PD (1998) The genus Eupodes Koch (Acari: Prostigmata: Eupodidae) from southern Africa with a redescription of E. parafusifer Meyer & Ryke and descriptions of two new species. Afr Entomol 6:275–288

Orlowska E, Orlowski D et al (2011) Role of mycorrhizal colonization in plant establishment on an alkaline gold mine tailing. Int J Phytoremediation 13(2):185–205. https://doi.org/10.1080/15226514.2010.495148

Otomo PV, Wepener V, Maboeta MS (2014) Single and mixture toxicity of gold nanoparticles and gold (III) to *Enchytraeus buchholzi* (Oligochaeta). Appl Soil Ecol 84:231–234

Otomo PV, Otomo LV, Bezuidenhout CC, Maboeta MS (2016) Preliminary evidence of differences in cadmium tolerance in metal-free stocks of the standard earthworm test species *Eisenia andrei* (Oligochaeta). Ecotoxicology 25(6):1119–1125

Owojori OJ, Reinecke AJ, Rozanov AB (2009a) Role of clay on partitioning, uptake and toxicity of zinc in the earthworm *Eisenia fetida*. Ecotoxicol Environ Saf 72:99–107

Owojori OJ, Reinecke AJ, Voua-Otomo P, Reinecke SA (2009b) Comparative study of the effects of salinity on life-cycle parameters of four soil dwelling species (*Folsomia candida, Enchytraeus doerjesi*, *Eisenia fetida* and *Aporrectodea caliginosa*). Pedobiologia 52:351–360

Owojori OJ, Reinecke AJ, Rozanov AB (2009c) The combined stress effects of salinity and copper on the earthworm *Eisenia fetida*. Appl Soil Ecol 41:277–285

Oyedele JA, Shimboyo S (2013) Distribution of radionuclides and radiation hazard assessment in soils of southern Namibia, Southern Africa. Radiat Prot Dosim 156(3):343–348. https://doi.org/10.1093/rpd/nct081

Pakade V, Cukrowska E et al (2013) Metal and flavonol contents of *Moringa oleifera* grown in South Africa. S Afr J Sci 109(3/4):10–15

Peakall DB (1992) Animal biomarkers as pollution indicators. Chapman & Hall, London

Picker MD, Hoffman MT, Leverton B (2007) Density of Microhodotermes viator (Hodotermitidae) mounds in southern Africa in relation to rainfall and vegetative productivity gradients. J Zool 271:37–44

Pickford GE (1937) A monograph of the Acanthodriline earthworms of South Africa. Priv. print. for the author by the Bournemouth Guardian

Plisko JD (1992a) Oligochaeta: Microchaetidae. Ann Natal Mus 37:295–307

Plisko JD (1992b) The Microchaetidae of Natal, with description of new species of Microchaetus Rapp and Tritogenia Kinberg, and the new genus Proandricus (Oligochaeta). Ann Natal Mus 33(2):337–378

Plisko JD (1995) New data on the biosystematics and distribution of Microchaetus natalensis (Kinberg, 1867) in north-eastern South Africa (Oligochaeta: Microchaetidae). Ann Natal Mus 36:281–291

Plisko JD (1996a) Michalakus, a remarkable new genus of microchaetid earthworm from South Africa (Oligochaeta: Microchaetidae). Ann Natal Mus 37:287–293

Plisko JD (1996b) Six new earthworm species of the southern African genus *Proandricus*. Ann Natal Mus 37:295–307

Plisko JD (1999) Designation of lectotypes for Microchaetus microchaetus (Rapp, 1849) and Microchaetus rappi Beddard, 1886, and historical perspectives on these species (Oligochaeta: Microchaetidae). Ann Natal Mus 40:269–276

Plisko JD (2001) Notes on the occurrence of the introduced earthworm Pontoscolex corethrurus (Müller, 1857) in South Africa (Oligochaeta: Glossoscolecidae). Afr Invertebr 42:323–334

Plisko JD (2002) Nine new earthworm species of Proandricus Plisko, 1992 from South Africa and Lesotho (Oligochaeta: Microchaetidae). Afr Invertebr 43:183–203
Plisko JD (2003) Eleven new South African earthworms (Oligochaeta: Microchaetidae) with new information on some known species, and an inventory of the microchaetids of KwaZulu-Natal. Afr Invertebr 44(2):279–325
Plisko JD (2004) Review of the balantine genus Udeina Michaelsen, 1910 with descriptions of six new species in South Africa (Oligochaeta: Acanthodrilidae, Acanthodrilinae). Afr Invertebr 45:287–313
Plisko JD (2006) A systematic reassessment of the genus Microchaetus Rapp, 1849: its amended definition, reinstatement of Geogenia Kinberg, 1867, and erection of a new genus Kazimierzus (Oligochaeta: Microchaetidae). Afr Invertebr 47:31–56
Plisko JD (2010) Megadrile earthworm taxa introduced to South African soils (Oligochaeta: Acanthodrilidae, Eudrilidae, Glossoscolecidae, Lumbricidae, Megascolecidae, Ocnerodrilidae). Afr Invertebr 51(2):289–312
Plisko JD (2012) Notes on the status of the family Microchaetidae: (Oligochaeta). Zool Middle East 58(sup 4):47–58
Plisko JD (2013) A new family Tritogeniidae for the genera Tritogenia and Michalakus, earlier accredited to the composite Microchaetidae (Annelida: Oligochaeta). Afr Invertebr 54(1):69–92
Posthuma L, Suter GW II, Traas TP (eds) (2002) Species sensitivity distribution in ecotoxicology. Lewis, Boca Raton
Rauch S, Fatoki OS (2013) Anthropogenic platinum enrichment in the vicinity of mines in the bushveld igneous complex, South Africa. Water Air Soil Pollut 224(1):1395. https://doi.org/10.1007/s11270-012-1395-y
Rauwane ME, Marais LM, Oladipo OG, Bezuidenhout CC, Maboeta MS (2018) Spatio-seasonal dynamics of bacteria and metal composition of a platinum mine tailings dam, Rustenburg, South Africa. S Afr J Plant Soil 35(2):129–135
Reinecke AJ, Ackerman D (1977) New earthworm species (Octochaetinae) from the northeastern Transvaal, South Africa. Wet Bydraes PU vir CHO Reeks B Nat 90:1–12
Reinecke AJ, Reinecke SA, Lambrechts H (1997) Uptake and toxicity of copper and zinc for the African earthworm, *Eudrilus eugeniae* (Oligochaeta). Biol Fertil Soils 24(1):27–31
Reinecke AJ, Reinecke SA, Maboeta MS (2001) Cocoon production and viability as endpoints in toxicity testing of heavy metals with three earthworm species. Pedobiologia 45(1):61–68
Reinecke AJ, Viljoen SA (1988) Reproduction of the African earthworm, *Eudrilus eugeniae* (Oligochaeta) – cocoons. Biol Fertil Soils 7(1):23–27
Reinecke AJ, Alberts JN (1994) Earthworm research in southern Africa since W. Michaelsen, with emphasis on the utilization of the earthworm (Eisenia fetida) as a protein source. Mitteilungen aus dem Naturhistorischen Museum in Hamburg 89(2):23–36
Reinecke AJ, Ljungström PO (1969) An ecological study of the earthworms from banks of the Mooi river in Potchefstroom, South Africa. Pedobiologia 9:106–111
Reinecke SA, Reinecke AJ (1997) The influence of lead and manganese on spermatozoa of *Eisenia fetida* (Oligocheata). Soil Biol Biochem 29:737–742
Reinecke AJ, Reinecke SA (1999a) Influence of exposure history to lead on lysosomal response in *Eisenia fetida* (Oligochaeta). Ecotoxicol Environ Saf 55:30–37
Reinecke SA, Reinecke AJ (1999b) Lysosomal response of earthworm coelomocytes induced by long-term exposure to heavy metals. Pedobiologia 43:585–593
Reinecke AJ, Reinecke SA (2003) The influence of exposure history to lead on lysosomal response in *Eisenia fetida*. Ecotoxicol Environ Saf 55:30–37
Reinecke SA, Reinecke AJ (2004) The comet assay as biomarker of heavy metal genotoxicity in earthworms. Arch Environ Contam Toxicol 46:208–215
Reinecke SA, Reinecke AJ (2007a) Biomarker response and biomass change of earthworms exposed to chlorpyrifos in microcosms. Ecotoxicol Environ Saf 66:92–101
Reinecke SA, Reinecke AJ (2007b) Impact of organophosphate pesticides in orchards on earthworms in the Western cape, South Africa. Ecotoxicol Environ Saf 66:244–251

Reinecke AJ, Ryke PAJ (1969) A new species of the genus Geogenia (Microchaetidae, Oligochaeta) from Lesotho, with notes on two exotic earthworms. Rev Ecol Biol Sol 6(4):515–523

Reinecke AJ, van Wyk M, Reinecke SA (2016) The influence of soil characteristics on the toxicity of oil refinery waste for the springtail Folsomia candida (Collembola). Bull Environ Contam Toxicol 96(6):804–809

Reinecke AJ, Visser FA (1980) The influence of agricultural land use practices on the population density of Allolobophora trapezoides and Eisenia rosea (Oligochaeta) in Southern Africa. In: Soil biology as related to land use practices. EPA, Washington, pp 310–324

Reinecke AJ, Venter JM, Viljoen SA (1990) The influence of feeding patterns on growth and reproduction of the vermicomposting earthworm Eisenia fetida (Oligochaeta). Biol Fert Soil 10:184–187

Reinecke AJ, Venter JM, Viljoen SA (1991) A comparison of the biology of Eisenia fetida and Eisenia andrei (Oligochaeta). Biol Fert Soil 11:295–300

Reinecke AJ, Viljoen SA, Saaymen RJ (1992) The suitability of Eudrilus eugeniae, Perionyx excavatus and Eisenia fetida (Oligochaeta) for vermicomposting in southern Africa in terms of their temperature requirements. Soil Biol Biochem 24(12):1295–1307

Reinecke SA, Reinecke AJ, Froneman MI (1995) The effects of dieldrin on the sperm ultrastructure of the earthworm, *Eudrilus eugeniae* (Oligochaeta). Environ Toxicol Chem 14:961–965

Reinecke SA, Helling B, Reinecke AJ (2002) Lysosomal response of earthworm *Eisenia fetida* coelomocytes exposed to the fungicide copper oxychloride and relation to life-cycle parameters. Environ Toxicol Chem 21:1026–1031

Reinecke AJ, Reinecke SA, Maboeta MS, Odendaal JP, Snyman RG (2007) Incorporating biomarkers in ecological risk assessment of chemical contaminants in soil. S Afr J Sci Technol 26:121–137

Reinecke AJ, Albertus RMC, Reinecke SA, Larink O, Griffiths CL (2008) The effects of organic and conventional management practices on feeding activity of soil organisms in vineyards. Afr Zool 43(1):66–74

Republic of South Africa (1996) Constitution of the Republic of South Africa, Act 108 of 1996. Government Gazette, Pretoria, p 378

Reynolds JW, Reinecke AJ (1976) A preliminary survey of the earthworms of the Kruger National Park, South Africa (Oligochaeta: Glossoscolecidae, Megascolecidae and Octochaetidae). Wet Bydraes PU vir CHO Reeks B Natuurwet 89:1–19

Rösner T, Van Schalkwyk A (2000) The environmental impact of gold mine tailings footprints in the Johannesburg region, South Africa. Bull Eng Geol Environ 59(2):137–148

Rossouw JM (2005) The use of different ecosystem components as indicators of ecosystem development during platinum mine tailings rehabilitation. MSc thesis. North-West University, Mahikeng

Ruhberg H, Hamer ML (2005) A new species of Opisthopatus Purcell, 1899 (Onychophora: Peripatopsidae) from KwaZulu-Natal, South Africa. Zootaxa 1039:27–38

Ryke PAJ, Loots GC (1967) The composition of the microarthropod fauna in South African soils. In: Satchell JE, Graff O (eds) Progress in soil biology. North-Holland Publ, Amsterdam

Schoeman JL, Van Deventer PW (2004) Soils and the environment: the past 25 years. S Afr J Plant Soil 21(5):369–387

Shaikh M, Moleele N et al (2006) Soil heavy metal concentration patterns at two speed zones along the Gaborone-Tlokweng Border Post Highway, Southeast Botswana. J Appl Sci Environ Manag 10(2):135–143

Simpson G, Berchner M (2017) Water-energy nexus-measuring integration: towards a water-energy-food nexus index. Water Wheel 16(1):22–23

Skaife SH (1955) THE ARGENTINE ANT: Iridomyrmex humilis Mayr. Trans R Soc S Afr 34(3):355–377

Snyman RG, Odendaal JP (2009) Effect of cadmium on haemocyte viability of the woodlouse *Porcellio laevis* (Isopoda; Crustacea). Bull Environ Contam Toxicol 83:525–529

Snyman RG, Reinecke SA, Reinecke AJ (2000) Haemocytic lysosome response in the snail *Helix aspersa*. Arch Environ Contam Toxicol 39:480–485

Snyman RG, Reinecke AJ, Reinecke SA (2002) Field application of a lysosomal assay as biomarker of copper oxychloride exposure in the snail *Helix aspersa*. Bull Environ Contam Toxicol 69:117–122

Somerset V, Horst CVD, et al (2012) Development of analytical sensors for the identification and quantification of metals in environmental samples. WRC Report; 2012. (2013/1/12): xv + 133 pp. Many ref. Gezina, Water Research Commission

South Africa (2014) National norms and standards for the remediation of contaminated land and soil quality in the Republic of South Africa. Government Gazette, Pretoria, p 37603

Spruyt A, Buck MT et al (2014) Arbuscular mycorrhiza (AM) status of rehabilitation plants of mine wastes in South Africa and determination of AM fungal diversity by analysis of the small subunit rRNA gene sequences. S Afr J Bot 94:231–237. https://doi.org/10.1016/j.sajb.2014.07.006

Stenersen J (1979) Action of pesticides on earthworms. Part I: The toxicity of cholinesterase-inhibiting insecticides to earthworms as evaluated by laboratory tests. Pestic Sci 10(1):66–74

Stenersen J (1980) Esterases of earthworms. part 1: characterisation of cholinesterases in *Eisenia foetida* (Savigny) by substrates and inhibitors. Comp Biochem Physiol 66C:37–44

Straker CJ, Weiersbye IM et al (2007) Arbuscular mycorrhiza status of gold and uranium tailings and surrounding soils of South Africa's deep level gold mines: I. Root colonization and spore levels. S Afr J Bot 73(2):218–225. https://doi.org/10.1016/j.sajb.2006.12.006

Street RA, Kulkarni MG, Stirk WA, Southway C, Abdillahi HS, Chinsamy M, Van Staden J (2009) Effect of cadmium uptake and accumulation on growth and antibacterial activity of *Merwilla plumbea* – an extensively used medicinal plant in South Africa. S Afr J Bot 75(3):611–616

Svendson C, Weeks JM (1997a) Relevance and applicability of a simple earthworm biomarker of copper exposure. Links to ecological effects in a laboratory study with *Eisenia andrei*. Ecotoxicol Environ Saf 36:72–79

Svendson C, Weeks JM (1997b) Relevance and applicability of a simple earthworm biomarker of copper exposure. Validation and applicability under field conditions in a mesocosm experiment with *Lumbricus rubellus*. Ecotoxicol Environ Saf 36:80–88

Svendson C, Meharg AA, Freestone P, Weeks JM (1996) Use of an earthworm lysosomal biomarker for the ecological assessment of pollution from an industrial plastics fire. Appl Soil Ecol 3:99–107

Theron PD, Meyer KP, Ryke PAJ (1969) Two new genera of the family Paratydeidae (Acari: Prostigmata) from South African soils. Acarologia 11:697–710

Traveset A, Richardson DM (2006) Biological invasions as disruptors of plant reproductive mutualisms. Trends Ecol Evol 21(4):208–216

Truhaut R (1975) Ecotoxicology: objectives, principles and perspectives. Ecotoxicol Environ Saf 1:151–173

Turnau K, Mesjasz-Przybylowicz J (2003) Arbuscular mycorrhiza of *Berkheya coddii* and other Ni-hyperaccumulating members of Asteraceae from ultramafic soils in South Africa. Mycorrhiza 13(4):185–190

Turnau K, Przybylowicz WJ et al (2013) Mycorrhizal fungi modify element distribution in gametophytes and sporophytes of a fern Pellaeaviridis from metaliferous soils. Chemosphere 92(9):1267–1273. https://doi.org/10.1016/j.chemosphere.2013.04.062

Ueckermann EA, Smith Meyer MK (1988) South African Acari. IV. Some Mites of the Addo Elephant National Park. Koedoe 31(1):31–51

United Nations Environment Programme (UNEP) (1999) Annual report. Evaluation and Oversight Unit, New York

United Nations Environment Programme (UNEP) (2013) Global environment outlook 2000, vol 1. Routledge, New York

Van Coller-Myburgh C, Rensburg LV, Maboeta M (2014) Utilizing earthworm and microbial assays to assess the ecotoxicity of chromium mine wastes. Appl Soil Ecol 83:258–265. https://doi.org/10.1016/j.apsoil.2013.09.004

Van Coller-Myburgh C, Van Rensburg L, Maboeta M (2015) Assessing the ecotoxicology of gold mine tailings utilizing earthworm and microbial assays. Water Air Soil Pollut 226(218):3–10

Van Den Berg RA, Ryke PAJ (1967) A systematic-ecological investigation of the acarofauna of the forest floor in Magoebaskloof (South Africa) with special reference to the Mesostigmata. Biol 6:157–234

Van Gestel CAM, Koolhaas JE, Hamers T, Van Hoppe M, Van Roovert M, Korsman C, Reinecke SA (2009) Effects of metal pollution on earthworm communities in a contaminated floodplain area: linking biomarker, community and functional responses. Environ Pollut 157:895–903

Van Nieuwenhuizen LC, Verster AJM, Horak IG, Krecek RC, Grimbeek JR (1994) The seasonal abundance of oribatid mites (Acari: Cryptostigmata) on an irrigated Kikuyu grass pasture. Exp Appl Acarol 18:73–86

Van Straalen NM, van Leeuwen CJ (2002) European history of species sensitivity distributions. In: Posthuma L, Suter GW II, Traas TP (eds) Species sensitivity distribution in ecotoxicology. Lewis, Boca Raton, pp 19–34

Vermeulen LA, Reinecke AJ, Reinecke SA (2001) Evaluation of the fungicide manganese-zinc ethylene bis (dithiocarbamate) (Mancozeb) for sublethal and acute toxicity to *Eisenia fetida* (Oligochaeta). Ecotoxicol Environ Saf 48(2):183–189

Vítková M, Ettler V, Hyks J, Astrup T, Kříbek B (2011a) Leaching of metals from copper smelter flue dust (Mufulira, Zambian Copperbelt). Appl Geochem 26:S263–S266

Vítková M, Ettler V, Mihaljevič M, Šebek O (2011b) Effect of sample preparation on contaminant leaching from copper smelting slag. J Hazard Mater 197:417–423

Voua Otomo P, Reinecke SA, Reinecke AJ (2013a) Combined effects of metal contamination and temperature on the potworm *Enchytraeus doerjesi* (Oligochaeta). J Appl Toxicol 33(12):1520–1524

Voua Otomo PV, Wahl J, Maboeta MS (2013b) The enchytraeid reproduction test (ERT): a potentially quick and affordable tool for the assessment of metal contaminated soils in emerging economies. Bull Environ Contam Toxicol 91(5):545–548. https://doi.org/10.1007/s00128-013-1092-6

Voua Otomo P, Reinecke SA (2010) Increased cytotoxic and genotoxic tolerance of *Eisenia fetida* (Oligochaeta) to cadmium after long-term exposure. Ecotoxicology 19:362–368

Voua Otomo P, Reinecke SA, Reinecke AJ (2014) Using the comet assay to assess the combined and separate genotoxic effects in *Eisenia andrei* (Oligochaeta) at different temperatures. Bull Environ Contam Toxicol 92:285–288

Wahl JJ, Theron PD, Maboeta MS (2012) Soil mesofauna as bioindicators to assess environmental disturbance at a platinum mine in South Africa. Ecotoxicol Environ Saf 86:250–260

Wardle DA, Bardgett RD, Klironomos JN, Setälä H, Van Der Putten WH, Wall DH (2004) Ecological linkages between aboveground and belowground biota. Science 304(5677):1629–1633

Webster R, Fey M (2011) Soils of South Africa. Exp Agric 47(2):411

Weiersbye IM, Witkowski ETF et al (2006) Floristic composition of gold and uranium tailings dams, and adjacent polluted areas, on South Africa's deep-level mines. Bothalia 36(1):101–127

Weldon CW, Daniels SR, Clusella-Trullas S, Chown SL (2013) Metabolic and water loss rates of two cryptic species in the African velvet worm genus Opisthopatus (Onychophora). J Comp Physiol B 183(3):323–332

Westcott M (2011) An evaluation of the germination and establishment of three selected coated grass species in different soil types for rehabilitation. MSc dissertation, North-West University, Potchefstroom Campus

Whitford WG, Eldridge DJ (2013) Effects of ants and termites on soil and geomorphological processes. Treatise Geomorphol 12:281–292

Witt ABR, Giliomee JH (2004) The impact of an invasive ant, Linepithema humile (Mayr), on the dispersal of Phylica pubescens Aiton seeds in South Africa. Afr Entomol 12(2):179

Zicsi A, Reinecke AJ (1992) Regenwürmer aus dem Krüger National Park in Süd Afrika (Oligochaeta: Eudrilidae). Acta Zool Acad Sci Hung 38(1–2):149–158

Phytoremediation of Heavy Metal-Contaminated Sites: Eco-environmental Concerns, Field Studies, Sustainability Issues, and Future Prospects

Gaurav Saxena, Diane Purchase, Sikandar I. Mulla, Ganesh Dattatraya Saratale, and Ram Naresh Bharagava

Contents

G. Saxena · R. N. Bharagava (✉)
Laboratory for Bioremediation and Metagenomics Research (LBMR), Department of Environmental Microbiology (DEM), Babasaheb Bhimrao Ambedkar University (A Central University), Lucknow, Uttar Pradesh, India
e-mail: gaurav_dem@bbau.ac.in; ramnaresh_dem@bbau.ac.in

D. Purchase
Department of Natural Sciences, Faculty of Science and Technology, Middlesex University, London, UK
e-mail: D.Purchase@mdx.ac.uk

S. I. Mulla
CAS Key Laboratory of Urban Pollutant Conversion, Institute of Urban Environment, Chinese Academy of Sciences, Xiamen, People's Republic of China
e-mail: sikandar@iue.ac.cn

G. D. Saratale
Department of Food Science and Biotechnology, Dongguk University-Seoul, Goyang-si, Gyeonggi-do, Republic of Korea
e-mail: gdsaratale@dongguk.edu

P. de Voogt (ed.), *Reviews of Environmental Contamination and Toxicology, Volume 249*, Reviews of Environmental Contamination and Toxicology Volume 249,
https://doi.org/10.1007/398_2019_24

Abbreviations

ABC	ATP-binding cassettes
ACC	1-Aminocyclopropane-1-carboxylate deaminase
Ag	Silver
AMF	Arbuscular mycorrhizal fungi
As	Arsenic
ATSDR	Agency for Toxic Substances and Disease Registry
BCF	Bioconcentration factor
CA	Citric acid
CAXs	Cation exchangers
Cd	Cadmium
CDF	Cation diffusion facilitators
Co	Cobalt
COPTs	Copper transporters
Cr	Chromium
Cu	Copper
CWs	Constructed wetlands
DDT	Dichlorodiphenyltrichloroethane
DIM	Daily intake of metal
DTPA	Diethylenetriaminepentaacetic acid
DW	Dry weight
EDDHA	Ethylenediamine-di(o-hydroxyphenylacetic acid)
EDDS	Ethylenediamine-N,N′-disuccinic acid
EDTA	Ethylenediaminetetraacetic acid
EGTA	Ethyleneglycolbis(b-aminoethyl ether),N,N,N0,N-tetraacetic acid
EKF	Electrokinetic field
ETCs	Evapotranspiration caps
EU	European Union
Fe	Iron
GEPs	Genetically engineered plants
GSH	Glutathione
HDL	High-density lipoprotein

HEDTA	N-Hydroxyethylenediaminetriacetic acid
Hg	Mercury
HMA	Heavy-metal ATPase
HMs	Heavy metals
IAA	Indole acetic acid
IARC	International Agency for Research on Cancer
LCA	Life cycle assessment
LDL	Low-density lipoprotein
MFC	Microbial fuel cell
MTPs	Membrane transporter proteins
MTs	Metallothioneins
Ni	Nickel
NTA	Nitrilotriacetic acid
PAHs	Polyaromatic hydrocarbons
Pb	Lead
PCBs	Polychlorinated biphenyl
PCs	Phytochelatins
PGPR	Plant growth-promoting rhizobacteria
PMFC	Plant-microbial fuel cell
PMIs	Plant-microbe interactions
ROS	Reactive oxygen species
SBMs	Soil beneficial microorganisms
TCE	Trichloroethylene
TF	Translocation factor
US EPA	US Environmental Protection Agency
WHO	World Health Organization
ZIP	Zinc-iron permease

1 Introduction

Environmental pollution is of serious ecological concern worldwide with a continually rising public outcry to ensure the safest and healthiest environment. A variety of organic and inorganic pollutants have been reported to cause environmental pollution and severe health hazards in living beings (Maszenan et al. 2011; Saxena and Bharagava 2017). Among them, heavy metals (HMs) are highly notorious pollutants due to their high abundance and non-biodegradable and persistent nature in the environment. Hence, they cause soil/water pollution and toxic, genotoxic, teratogenic, and mutagenic effects in living beings (Dixit et al. 2015; Sarwar et al. 2017). They also cause endocrine disruption and neurological disorders even at low concentration (Yadav 2010; Maszenan et al. 2011; Dixit et al. 2015; Sarwar et al. 2017). Any naturally occurring metal/metalloid having an atomic number greater than 20 and elemental density greater than 5 g cm^{-3} is termed as HM. They include copper (Cu), cadmium (Cd), chromium (Cr), cobalt (Co), zinc (Zn), iron (Fe), nickel (Ni), mercury (Hg), lead (Pb), arsenic (As), silver (Ag), and platinum group elements (Ali et al. 2013; Ali and Khan 2018a). Among them, Cd, As, Hg, and Pb don't have any biological function in the body and thus are nonessential elements. They can cause severe health hazards and are listed as priority pollutants by many environmental protection agencies worldwide (Jaishankar et al. 2014; Dixit et al. 2015; Sarwar et al. 2017). Therefore, the removal of HMs from the contaminated matrix is an urgent need to safeguard the environment and human health. Phytoremediation has been identified as an emerging, low-cost, and eco-sustainable solution for HM pollution prevention and control. It is the most suitable alternative to conventional physicochemical remediation technologies, which are highly expensive and technically more suited to small areas, create secondary pollution and deteriorate soil fertility, and thus adversely affect agroecosystem (Ali et al. 2013; Chandra et al. 2015; Mahar et al. 2016; Muthusaravanan et al. 2018).

Phytoremediation is the engineered use of green plants with associated soil beneficial microbes to remove toxic pollutants via degradation and detoxification mechanisms from contaminated soil and water/wastewaters (Bharagava et al. 2017a; Mukhopadhyay and Maiti 2010; Ali et al. 2013). It is an eco-friendly, non-intrusive, and aesthetically pleasing remediation technology that removes metal pollutants from the contaminated sites (Lee 2013; Chandra et al. 2015; Chirakkara et al. 2016). It can be commercialized, and income can be generated, if metals removed from contaminated sites could be used to extract usable form of economically viable metals (i.e., phytomining) (Chandra et al. 2015; Mahar et al. 2016). In addition, energy can be generated through the burning of plant biomass, and land restoration could be achieved for sustainable agricultural development or general habitation (Stephenson and Black 2014; Mahar et al. 2016). The rationale, mechanisms, and economic feasibility of phytoremediation have been discussed elsewhere (Ali et al. 2013; Wan et al. 2016; Sarwar et al. 2017). However, extensive research is currently underway to testify the phytoremediation potential of hyperaccumulating plants at field scale for the treatment and management of HM-contaminated sites.

The deadly poisonous and indestructible nature of HMs is mainly responsible for the eco-toxicity and health hazards. Thus, an eco-friendly solution (i.e., phytoremediation) is required for the treatment and management of HM-contaminated sites. This paper aims to provide a comprehensive review on the following areas of phytoremediation: (a) environmental pollution and toxicity profile of HMs; (b) conventional and novel phytoremediation approaches and their role in environmental management with merits and demerits; (c) field studies and sustainability issues associated with phytoremediation of HM-contaminated sites; (d) plant-microbe interactions (PMIs) and their role in enhanced phytoremediation; (e) challenges and opportunities for valorization of plant biomass in biofuel/bioenergy production; (f) challenges in transgenic approaches to modify the hyperaccumulating plants (designer plants) and associated microbes (engineered bacteria); and (g) the knowledge gaps and potential areas for further research in the phytoremediation of HM-contaminated sites.

2 Sources of Heavy Metal Contamination and Toxicity in Environment

HMs can be introduced into the environment either by natural or anthropogenic processes. Natural processes are geological activities, for instance, mineral weathering, erosion, volcanic eruptions, and continental dust. Anthropogenic activities include industrial operations such as mining, smelting, electroplating, and industrial effluent discharge as well as agricultural practices like use of pesticides and phosphate fertilizers and release of agricultural wastes (Ali et al. 2013; Mahar et al. 2016; Antoniadis et al. 2017). Industrial activities are the major source of HM pollution (water and soil) in the environment. If HMs enter the food chain, they may bioaccumulate and/or biomagnify at higher trophic levels resulting in severe health threats and thus are of serious ecotoxicological concern.

The indiscriminate discharge of toxic metal-rich industrial effluents is one of the major sources of environmental pollution. The effluent discharged from metal-based industries, especially leather industries (Cr used in leather tanning), causes serious soil and water pollution, and hence, its treatment and management is a key challenge to pollution control authorities (Sahu et al. 2007; Saxena et al. 2016). A high concentration of HMs has been reported in sediments of river Ganga and its tributaries receiving Cr-loaded tannery effluent (Beg and Ali 2008). In addition, HMs beyond the permissible limits also deteriorate water quality and make it unfit for drinking and irrigation purpose (Nazeer et al. 2014). The effluent released from electroplating and distillery industries also constitutes a highly rich source of HMs and, hence, is considered as hazardous to living beings (Venkateswaran et al. 2007; Chandra et al. 2008). Furthermore, effluent released from domestic activities is also responsible for HM pollution and thus is of serious ecotoxicological concerns (Bhardwaj et al. 2017).

In an aquatic ecosystem, HMs adversely affect gamete production, sperm quality, and embryonic development, delay hatching, and cause physical deformities in fishes which ultimately leads to the death of newly hatched larvae (Segura et al. 2006; Jezierska et al. 2009; Fatima et al. 2014). HMs also cause endocrine disruption, oxidative stress, and genotoxicity in fishes (Jezierska et al. 2009; Łuszczek-Trojnar et al. 2014; Javed et al. 2016). Further, HMs also cause a reduction in hematological parameters and glycogen reserve and thus make the fishes weak, anemic, and vulnerable to diseases (Javed and Usmani 2015).

The soil is a nonrenewable resource for sustainable agriculture and acts as a major sink for HMs. The contamination of agricultural soil with toxic metals affects its physicochemical and biological properties and reduces land usability for agricultural farming leading to food insecurity and thus creates land tenure problems (Wuana and Okieimen 2011). Moreover, the coexistence and persistence of HMs in soil are also responsible for the entry of toxic metals into the food chain and thus lead to severe health hazards in living beings (Khan et al. 2008) (Table 1).

HMs inhibit several microbial metabolic processes such as respiration, denitrification, and enzymatic activity and hence retard the bioremediation processes

Table 1 Sources of contamination and toxicity profile of heavy metals

Heavy metal	Standard limit		Sources of contamination	Environmental hazards	Toxic effects	
	Water (mg/L)[a]	Soil (mg kg^{-1})[b]			Plants	Humans/animals
Arsenic (As)	0.01	1–50	Pesticides and wood preservatives	Water and soil pollution	Analog of phosphate (P) and hence compete for the uptake of carriers (P/As) in root plasmalemma and thus disrupt phosphate-dependent metabolism	Analog of phosphate and hence affects oxidative phosphorylation and ATP synthesis, the sensation of "pins and needles" in hands and feet
Cadmium (Cd)	0.003	0.01–0.7	Paints and pigments, plastic stabilizers, electroplating, incineration of cadmium-containing plastics, phosphate fertilizers	Water and soil pollution	Chlorosis; browning of root tips; reduced seed germination, growth, photosynthesis, water, nitrate, and nutrient uptake; and ATPase activity and Fe(II) deficiency, and ultimately death	Carcinogenic, mutagenic, teratogenic; endocrine disruptor, hypercalciuria, and itai-itai disease
Lead (Pb)	0.01	2–200	Aerial emission from combustion of leaded petrol, battery manufacture, herbicides, and insecticides	Water and soil pollution	Chlorosis, reduced seed germination, growth, biomass, photosynthesis, nutrient and water uptake and transport, alter membrane permeability, induce abnormal morphology, oxidative stress (ROS generation) in plants, and inhibit enzymatic activity at the cellular level by reacting with their sulfhydryl groups	Impaired development, reduced intelligence, short-term memory loss, insomnia, anorexia, encephalopathy, disabilities in learning and coordination problems, the risk of cardiovascular disease, foot drop/wrist drop (palsy), and nephropathy

Chromium (Cr)	0.05	1–1,000	Tanneries, steel industries, fly ash, dyes, and pigments	Water and soil pollution	Chlorosis, membrane damage, nutrient imbalance, wilting of tops, and root injury, reduced seed germination, growth and development, photosynthesis, and enzymatic activity	Highly toxic proven carcinogen as identified by IARC, WHO, ATSDR, and USEPA; hair loss, pulmonary fibrosis (lung scarring), lung cancer and damage to the kidney, circulatory and nerve tissues
Mercury (Hg)	0.001	0.01–0.3	Release from Au-Ag mining and coal combustion, surgical instruments, medical waste	Water and soil pollution	Interfere with mitochondrial activity and induce oxidative stress by triggering ROS generation and thus disrupt membrane lipids and cellular metabolism	Possible human carcinogen (methyl-Hg) as established by USEPA; anxiety; Minamata; autoimmune diseases; depression; difficulty with balance; drowsiness; fatigue; hair loss; insomnia; irritability; memory loss; recurrent infections; restlessness; vision disturbances; tremors; temper outbursts; ulcers and damage to the brain, kidney, and lungs; neurasthenia (neurotic disorder); and parageusia (metallic taste)
Copper (Cu)	2.00	2–100	Pesticides and fertilizers	Water and soil pollution	Chlorosis, reduced growth, induce stress (ROS generation) and thus disturb metabolic pathways and damage to macromolecules, exert cytotoxic effects, and ultimately injury to plants	Wilson's disease, brain and kidney damage, liver cirrhosis, chronic anemia, stomach and intestine irritation, and even death

(continued)

Table 1 (continued)

Heavy metal	Standard limit		Sources of contamination	Environmental hazards	Toxic effects	
	Water (mg/L)[a]	Soil (mg kg^{-1})[b]			Plants	Humans/animals
Nickel (Ni)	0.02	5–500	Industrial effluents, kitchen appliances, surgical instruments, steel alloys, automobile batteries	Water and soil pollution	Reduced seed germination, chlorosis, necrosis, nutrient imbalance, ion imbalance particularly $K+$, alteration in cell membrane functions, lipid composition, and H-ATPase activity of the plasma membrane	Hematotoxic, immunotoxic, neurotoxic, genotoxic, reproductive toxic, pulmonary toxic, nephrotoxic, and hepatotoxic; allergic dermatitis (itching); cancer of the lungs, nose, sinuses, throat, and stomach; hair loss and defects in infants; cardiovascular and musculoskeletal system

Adapted from Sarwar et al. (2017), Dixit et al. (2015), Jaishankar et al. (2014), Ali et al. (2013), and Yadav (2010)

IARC International Agency for Research on Cancer, *WHO* World Health Organization, *ATSDR* Agency for Toxic Substances and Disease Registry, *USEPA* US Environmental Protection Agency, *ROS* reactive oxygen species, *LDL* low-density lipoprotein, *HDL* high-density lipoprotein

[a]Standard limit of heavy metals in drinking water according to WHO drinking water standard (1993) for standard setting and drinking water safety. Access online at https://www.lenntech.com/applications/drinking/standards/who-s-drinking-water-standards.htm

[b]Common range of heavy metals in soil according to Lindsay (1979). Access online at http://www.occeweb.com/og/metals-limits.pdf

(Zhuang et al. 2007; Sobolev and Begonia 2008). HMs also cause a reduction in the number of specific microbial populations and a shift in the microbial community structure. For instance, Ding et al. (2016) evaluated the effect of Cd and Cr on the microbial community structure in the rhizospheric soil of rice plant during a pot experiment. Results revealed that the relative abundance of a bacterial genus *Longilinea* was significantly higher in the control soil than in Cd- and Cr-treated soils, whereas the relative abundance of the genus *Pseudomonas* was significantly higher in the Cd-treated soils than in the Cr-treated and control soils. However, the relative abundance of the genus *Sulfuricurvum* was also significantly higher in the Cd-treated soil than in the Cr-treated and control soils, whereas the relative abundance of the genus *Bellilinea* was significantly higher in the Cr-treated soil than in the other treated soils. HMs also inhibit the cell division, transcription process, and denaturation of protein and adversely affect the cell membrane distribution in microbes (Jacob et al. 2018). Hexavalent chromium (Cr^{6+}) is also reported to cause DNA damage by exerting oxidative stress in soil bacteria and thus leads to genotoxic effects (Quievryn et al. 2003).

The irrigation of food crops in the agriculture field with water contaminated with toxic metal-rich industrial effluents is a common practice in many developing countries. It may provide a chance for the movement of potentially toxic metals from contaminated soil to edible crops, which ultimately reach into the human/animal body via consumption and thus render severe toxic effects. HMs affect various metal-sensitive enzymes in plants such as alcohol dehydrogenase, nitrogenase, nitrate reductase, amylase, and hydrolytic (phosphatase and ribonuclease) and carboxylating (phosphoenolpyruvate carboxylase and ribulose-1,5-bisphosphate carboxylase) enzymes (Nagajyoti et al. 2010; Yadav 2010). Hence, HMs disrupt several biochemical/physiological processes in plants such as seed germination, enzymatic activities, nitrogen metabolism, electron transport system, transpiration, CO_2 assimilation, antioxidant defense system, photosynthesis, photophosphorylation, cellular metabolism, nitrogen fixation, water balance, mineral nutrition, and cellular ionic homeostasis, which ultimately leads to plant's death (Yadav 2010; Lajayar et al. 2017). Irrigation of agricultural crops with heavy metal-loaded industrial effluents also disrupts several cytological processes in plants such as root growth and elongation, cell membrane permeability, mitotic activity, and the stability of genetic material and also creates chromosomal abnormalities (Nagajyoti et al. 2010; Yadav 2010). For example, the irrigation of agricultural crops with the HM-rich distillery and tannery effluent has been reported to cause a reduction in root/shoot growth and biomass, seed germination, and seedling growth and also induce chlorosis and photosynthetic impairment (Chandra et al. 2009; Bharagava et al. 2017b).

HMs may cause oxidative stress by forming reactive oxygen species (ROS), which disrupt the antioxidant's defense system and lead to cell damage in humans/animals and in extreme cases can be fatal (Jaishankar et al. 2014). For instance, hexavalent chromium (Cr^{6+}) has been reported to cause cancer in humans and damage cellular components during its reduction into trivalent chromium (Cr^{3+}), leading to the generation of free radicals that cause DNA damage (Mishra and

Bharagava 2016). Therefore, the remediation of HM-contaminated sites is utmost important for environmental safety.

3 Trophic Transfer of Toxic Heavy Metals and Its Consequences

Trophic transfer or biotransference is an ecological phenomenon by which a contaminant enters the food chain through uptake either from ambient abiotic environment (bioconcentration) or both ambient abiotic environment and organism's food/diet (bioaccumulation) and passage from one trophic level to the next higher trophic level (biomagnification) and consequently poses risks to human/animal health (Ali and Khan 2018a, b). The trophic transfer of toxic HMs from soil to plants to humans and organism's food to humans is depicted in Fig. 1. The primary route of HM entry into the food chain is through the soil-to-plant transfer mechanism. In the soil-to-plant transfer mechanism, HMs are transferred from soil to agricultural crops/vegetables that constitute a large source of human diet and thus may result in catastrophic health hazards (Table 1). According to a study, the daily intake of metal (DIM) was higher for vegetables grown on soils irrigated with HM-rich

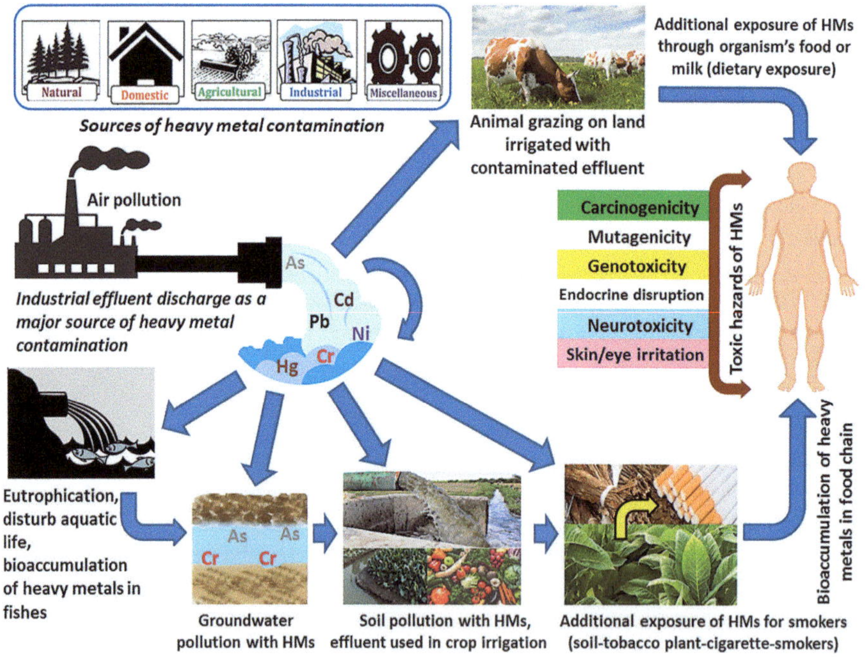

Fig. 1 Trophic transfer of toxic HMs from soil to plants to humans and organism's food to humans and their toxicity

wastewater compared to those of control soils (Jan et al. 2010). The consumption of fishes contaminated with toxic HMs also poses serious risks to human health. Hence, their dietary role has been questioned though they provide omega-3 fatty acids that have cardioprotective effects. In the 1950s, Minamata disease in Japan caused by the consumption of Hg-contaminated fishes by the local people is considered as one of the major environmental chemical disasters of the twentieth century (Ali and Khan 2017). Moreover, HM accumulation in invertebrates (due to varying feeding habits of insects), amphibians and reptiles (due to absorption through highly permeable skin), and birds (due to ingestion of contaminated food and water) also adversely affects their development, growth, health, feeding behavior, physiology, and reproduction (Ali and Khan 2018b). Thus, the trophic transfer, bioaccumulation, and biomagnification of toxic HMs in food chains have important implications for wildlife and human health. Further, more details on the subject can be found in a good review article published by Ali and Khan (2018a, b).

4 Phytoremediation Approaches for Environmental Cleanup

The engineered use of green plants with associated beneficial microbes to degrade/ detoxify pollutants from the contaminated medium (soil/water/wastewaters) is technically described as phytoremediation. The term "phytoremediation" is made up of two words, i.e., Greek head "*phyto*" (means plant) and Latin root "*remedium*" (means to correct or remove an evil). It can be applied for the eco-restoration of sites primarily contaminated with HMs, radionuclides, and various recalcitrant organic pollutants (Ali et al. 2013; Mahar et al. 2016). It comprises different phytotechniques for the amelioration of various pollutants using different mechanisms depending on their applications (Fig. 2). However, all the mechanisms cannot be applied for the remedy of all the pollutants. Different phytoremediation techniques such as phytoextraction, phytostabilization, phytovolatilization, rhizodegradation, phytodegradation, and rhizofiltration have been extensively discussed elsewhere (Ali et al. 2013; Chandra et al. 2015; Chirakkara et al. 2016; Sarwar et al. 2017). The definition, application, and bottlenecks of traditional phytoremediation techniques are summarized in Table 2.

Among the phytoremediation techniques, phytoextraction is a major mechanism of HM removal from contaminated sites. When green plants are used to remove metal pollutants from contaminated sites via root absorption and their sequestration/ concentration in the aboveground harvestable parts is technically described as phytoextraction (Vangronsveld et al. 2009; Mukhopadhyay and Maiti 2010; Mahar et al. 2016). It could be an economically viable technology when metals extracted from contaminated sites using plant biomass can be utilized as "bio-ore" to extract the functional and valuable metals, and this process is technically regarded as phytomining (Chandra et al. 2015). Thus, it can generate income and offer additional

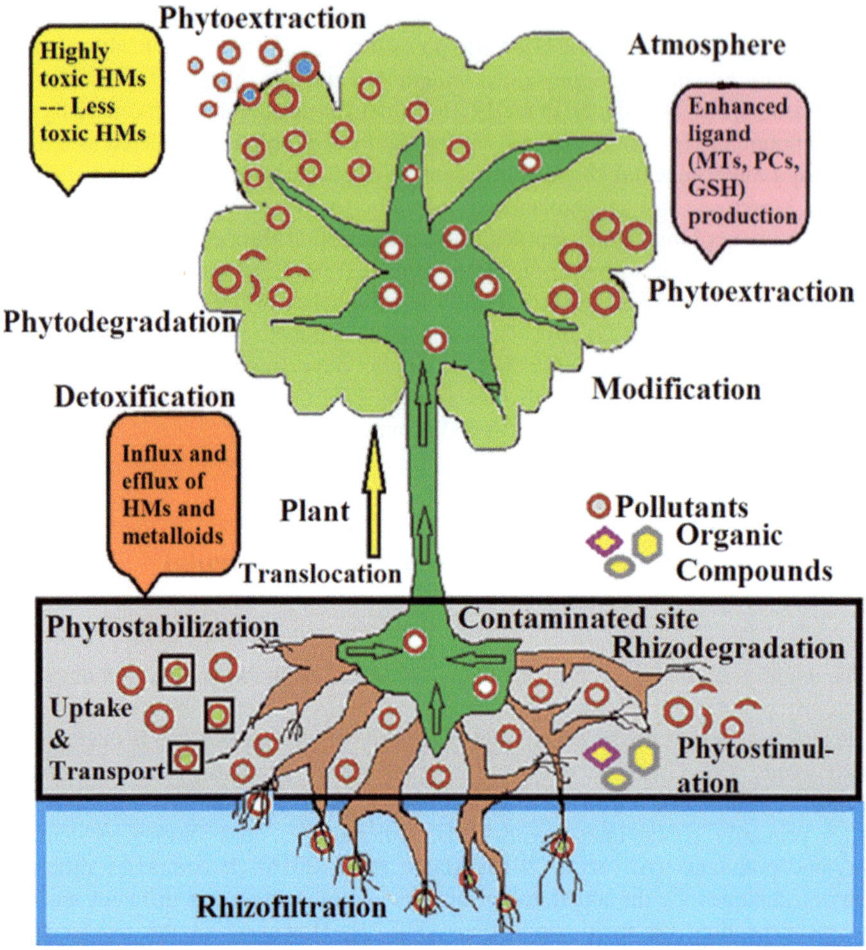

Fig. 2 A pictorial representation of different phytoremediation techniques

employment opportunity to the public (Sheoran et al. 2009; Chandra et al. 2015; Stephenson and Black 2014).

The phytoextraction efficiency of green plants primarily depends on the bioconcentration factor (BCF) and translocation factor (TF). BCF represents metal concentration in root/soil and denotes metal accumulation, whereas TF represents metal concentration in shoot/root and denotes metal translocation (Goel et al. 2009; Ali et al. 2013; Antoniadis et al. 2017). Plants with high biomass, fast growth rate, and high metal tolerance and accumulation are chiefly preferred for metal's phytoextraction (Mukhopadhyay and Maiti 2010; Lee 2013; Chandra et al. 2015). Phytoextraction is performed in two different ways (Ali et al. 2013): natural (plants accumulate metals under natural conditions) and induced or assisted (application of enhancers to increase metal accumulation in plants). Enhancers are used to increase

Table 2 Description of phytoremediation mechanisms and applications

Phytoremediation processes	Definition	Mechanism	Pollutants	Applicability	Benefits	Comments/issues
Phytoextraction or phytoaccumulation or phytoabsorption or phytosequestration	Plants remove metal pollutants from contaminated sites via plant's root absorption and sequester/concentrate in aboveground harvestable plant parts	Hyperaccumulation	Pb, Cd, Zn, Ni, Cu, Pb, radionuclides, pentachlorophenol, aliphatic compounds (short chained)	Contaminated soil/sites, water, wastewaters	Abundant biomass in short time, reduced soil erosion and cost-effective, wide application prospects	Slow process; contaminant concentration is important, and it depends on the depth of contamination, the risk of metal leaching, and thus groundwater pollution; require postharvest treatment for contaminated biomass volume reduction, metal recovery (i.e., phytomining), and bioenergy production
Phytofiltration or rhizofiltration	Plants concentrate and precipitate metal pollutants in low concentration from the aquatic environment in their roots	Rhizosphere accumulation	Pb, Cd, Zn, Ni, Cu, radionuclides (Cs, Sr, U), hydrophobic organics, and radionuclides	Contaminated water and wastewaters	Cleanup of polluted surface water, industrial wastewaters, and agricultural runoff	Plant roots act as filters for cleanup of polluted water/wastewaters; less generation of secondary waste and minimize the need of further disposal, if terrestrial plants are used due to

(continued)

Table 2 (continued)

Phytoremediation processes	Definition	Mechanism	Pollutants	Applicability	Benefits	Comments/issues
						high biomass as compared to aquatic plants (highly species specific); long-term maintenance depends on the type of contaminant and contamination depth
Phytostabilization or phytoimmobilization or phytotransformation	Plants immobilize or inactivate metal pollutants at their place involving absorption by roots, adsorption onto roots, and precipitation, complexation, and metal valence reduction in rhizosphere, e.g., reduction of Cr^{6+} to Cr^{3+}	Precipitation, complexation, and metal valence reduction	Pb, Cd, Zn, As, Cu, Cr, Se, U, polycyclic aromatic hydrocarbons (PAHs), polychlorinated biphenyl (PCBs), dioxins, furans, pentachlorophenol, DDT, dieldrin	Contaminated soil/sediments and sludge	Ecologically efficient, stabilization of contaminated medium without disposal of contaminated biomass, reduces soil erosion, applicable in the field and mine polluted areas	Not a permanent solution because plants only limit mobility/bioavailability of metal pollutants at the place, and hence the site cannot be used for plant growth
Phytovolatilization or phytoevaporation	Plants uptake metal pollutants through roots in low concentration, modify/transform them into less toxic form, and	Volatilization or evaporation by leaves	Chlorinated solvents like carbon tetrachloride, trichloroethylene, methylene chloride, tetrachloroethylene,	Contaminated wastewaters, soil, sediments, and sludges	Environmental cleanup without harvesting plants and biomass disposal	Suitable for Hg^{2+} removal; limited in case of Se, elemental Hg, and As due to their gaseous forms; most

	subsequently transpire/volatilize into the atmosphere through stomata		carbon tetrachloride, 1,1,1-trichloroethane, Hg (mercuric ion), Se			controversial because there may be a chance of staying toxic metals in air and thus no control over migration (air pollution); redeposition of pollutant back into the ecosystem by precipitation (elemental Hg)
Phytodegradation	Plants breakdown/convert highly toxic organic pollutants into less toxic forms through the action of enzymes secreted within plant tissues and released in the rhizosphere	Degradation in plant tissues	DDT, PAHs, bisphenol A, organophosphorus compounds	Contaminated soil, sediments, sludges, groundwater, surface water, and wastewaters	Biodegradation of various recalcitrant pollutants in the rhizosphere	Key enzymes for degradation are nitroreductase, dehalogenase, oxygenase, peroxidase, nitrilase, nitroreductase, and laccase; depends on factors such as concentration and composition, plant species, and soil conditions
Rhizodegradation or rhizoremediation or phytostimulation	Plants break down/convert highly toxic organic pollutants into less toxic forms through enzymatic activity of rhizospheric microorganisms	Degradation in rhizosphere	Atrazine, ammunition wastes, petroleum hydrocarbon, PCBs, PAHs, TCE, diesel fuel	Contaminated soil, sediments, sludges, groundwater, and wastewaters	Release of organic acid; rhizosphere enhances biodegradation; metabolic products are further utilized by microorganisms in rhizosphere	Depends on the action of root exudates and enzymes; increased activity of rhizosphere microbes

the phytoextraction efficiency and include chelators or soil amendments (Sarwar et al. 2017). Chelators are the organic and mineral acids that increase the bioavailability of insoluble or unavailable form of metals in soil making them available for plant uptake (Ali et al. 2013; Mahar et al. 2016), thus enhancing the phytoremediation efficiency by solving the low metal phytoavailability issue. Some commonly available chelators are EDTA, ethylenediaminetetraacetic acid; HEDTA, N-hydroxyethylenediaminetriacetic acid; DTPA, diethylenetriaminepentaacetic acid; EGTA, ethyleneglycolbis(b-aminoethyl ether),N,N,N0,N-tetraacetic acid; EDDHA, ethylenediamine-di(o-hydroxyphenylacetic acid); EDDS, ethylenediamine-N,N′-disuccinic acid; NTA, nitrilotriacetic acid; and CA, citric acid. The ability of chelators to enhance the metal accumulation in plants has been tested and reviewed by many workers (Xie et al. 2012; Ramamurthy and Memarian 2014; Sun et al. 2015; Chirakkara et al. 2016). Organic soil amendments are cheaper, eco-friendly, and non- or less toxic and degradable in nature. These help to minimize environmental pollution and reduce toxicity to remediating plants and ultimately enhance phytoremediation efficiency (Wiszniewska et al. 2016). Some specific kinds of organic soil amendments include agro- and industrial wastes (such as sugar beet residue, wheat, and rice straw, composted sewage sludge or molasses), biochar, compost, humic substances, plant extracts, and exudates and are of great significance in HM phytoremediation (Wiszniewska et al. 2016). The use of organic soil amendments to enhance the phytoremediation efficiency has been evaluated (Park et al. 2011; Paz-Ferreiro et al. 2014; Wiszniewska et al. 2016; Chirakkara et al. 2016; Reddy et al. 2017). The biotic and abiotic factors also affect the efficiency of phytoremediation (Fig. 3). Biotic factors include plant and root zone characteristics, whereas abiotic factors comprise pollutant and chelator characteristics, properties of the medium (e.g., soil), and climate conditions.

Further, the environmental risks associated with synthetic chelators such as low biodegradability of chelators, groundwater contamination due to leaching of highly water-soluble metal-chelator complexes to deeper layer of soil, slow decomposition of organic acids, and toxicity to remediating plants and soil beneficial microorganisms (SBMs) should also be considered before application (Vangronsveld et al. 2009; Stephenson and Black 2014; Mahar et al. 2016). However, the selection of chelators with optimum dose and application time could help to minimize the associated environmental risks and toxic effects in remediating plants.

5 Hyperaccumulating Plants for Phytoremediation of Heavy Metals

5.1 Classification of Metallophytes

Plants that tolerate and survive in soil containing an exceptionally high concentration of HMs without suffering toxicity are termed as metallophytes. The soil where ores

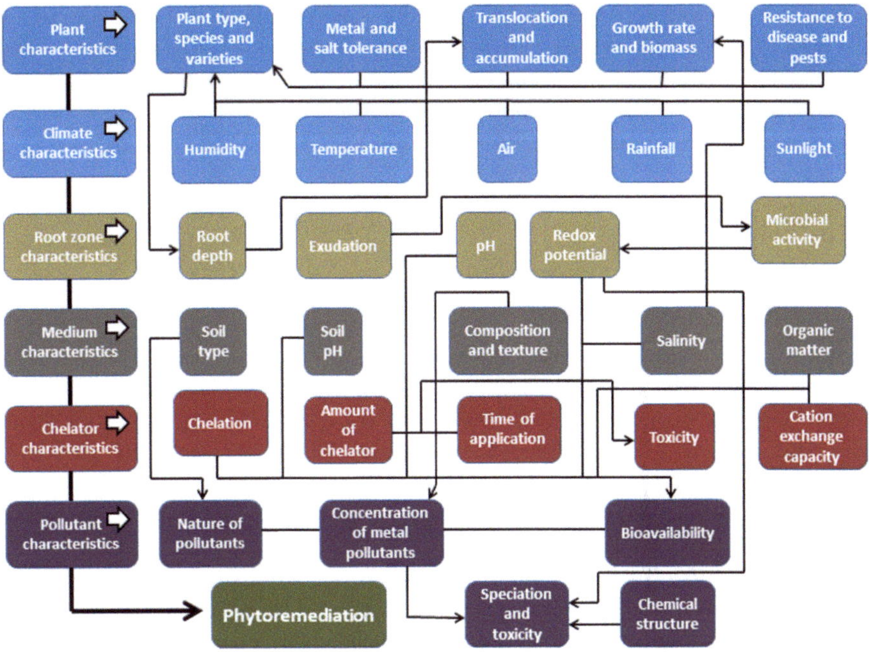

Fig. 3 Relationships among the factors affecting phytoremediation efficiency

are outcropping is termed as metalliferous or orogenic soil. According to growth potential in HM-contaminated sites, metallophytes can be grouped into the following categories: (a) metal excluders (those that accumulate metals in their roots but restrict transport and entry into their aerial parts possibly by altering cell membrane permeability and changing cell wall metal-binding capacity via modulating ionic channels, ion pump activity, and activation of new ionic conductance or exudating more chelating substances in soil); (b) metal indicators (those that actively accumulate metals in their aerial parts by releasing intracellular metal-binding chemicals, i.e., chelators, or altering the pattern of metal compartmentalization by storing them in nonsensitive plant parts such as vacuoles and cell wall and generally reflect metal concentration in soil); and (c) metal accumulators (those that actively accumulate exceedingly large concentration of metals from the soil in the aboveground plant parts, especially leaves with no symptoms of phytotoxicity) (Mukhopadhyay and Maiti 2010; Chandra et al. 2015; Antoniadis et al. 2017). The use of metallophytes alone or in combination with microorganisms is an excellent strategy for the phytoremediation and HM pollution prevention and control.

To date, several metallophytes have been identified and used in the phytoremediation of HM-contaminated sites. Some specific examples include *Pteris vittata*, which can accumulate Cr and As up to 35,303 and 20,707 mg kg^{-1} dry weight (DW), respectively (Kalve et al. 2011); *Alyssum murale*, which can accumulate Ni in range of 4,730–20,100 mg kg^{-1} DW (Bani et al. 2010); *Tagetes minuta*, which can accumulate As up to 380.5 mg kg^{-1} DW (Salazar and Pignata 2014); *Eleocharis acicularis*, which can accumulate Zn up to 11,200 mg kg^{-1} DW (Sakakibara et al. 2011); *Corrigiola telephiifolia*, which can accumulate As up to 2,110 mg kg^{-1} DW (Garcia-Salgado et al. 2012); and *Noccaea caerulescens*, which can accumulate Pb in range of 1,700–2,300 mg kg^{-1} DW (Dinh et al. 2018).

5.2 Selection Criteria for Hyperaccumulating Plants for Phytoremediation

Plants (woody/herbaceous) that accumulate high metal concentration in their shoot (100–1,000-fold higher than those found in non-hyperaccumulating species) without any visible symptoms are regarded as hyperaccumulators (represented by <0.2% of angiosperms), and the overall process is termed as hyperaccumulation (Ghosh and Singh 2005; Mukhopadhyay and Maiti 2010; Lee 2013). An ideal hyperaccumulator plant should accumulate at least 100 mg kg^{-1} (0.01% dry wt.) of As and Cd; 1,000 mg kg^{-1} (0.1% dry wt.) of Co, Cu, Cr, Ni, Se, and Pb; and 10,000 mg kg^{-1} (1% dry wt.) of Zn and Mn (Reeves and Baker 2000). Metal hyperaccumulating plants are chiefly preferred for phytoremediation. However, some hyperaccumulating plants are not successfully applied because of several reasons: (a) low biomass, (b) slow growth rate, (c) metals (such as Ni, Zn, and Cu) under scope of phytoextraction that are not the priority pollutants, and (d) agronomic practices and crop protection measures for their cultivation and protection that have not been developed (Goel et al. 2009; Marques et al. 2009; Mahar et al. 2016; Yadava et al. 2018).

Serpentine/ultramafic soils (most abundant metalliferous soil on earth) are the most important natural resource in the screening of hyperaccumulating plants especially for metals like Ni and/or Zn. Plant species inhabiting such soils are called as "serpentinophytes." Serpentinophytes are of great significance in phytoremediation due to their specific adaptation (might be confined to the physiological state) to such high HM-rich soil. Several plant species like *Crotalaria micans*, *Leucaena leucocephala*, *Bidens pilosa*, *Pueraria lobata*, and *Conyza canadensis* have been reported to remove Ni from a serpentine site for successful phytoremediation applications in Taiwan (Ho et al. 2013). Similarly, many research reports have been published on the screening of hyperaccumulating plants for metals like Zn and Ni from serpentine soils (Ho et al. 2013; Tomović et al. 2013; Salihaj et al. 2016; Bini et al. 2017). According to Bini et al. (2017), *Alyssum bertolonii* accumulated an exceptionally high concentration of Ni (i.e., 2,118 mg kg^{-1} in aerial parts) from serpentine soils of Tuscany (Central Italy). They also suggested the use of *Alyssum bertoloni* in the remediation of Ni-contaminated soils with both the phytoextraction and phytostabilization and in phytomining as well. Thus, serpentinophytes are the key metal hyperaccumulators and amicable for the phytoremediation and phytomining. However, the ecological significance of metal hyperaccumulation in plants is still unclear. It has been suggested that the hyperaccumulation in plants might result from the development of high metal tolerance, drought resistance, unintended metal absorption, competition with other metal-tolerant plants, and defense mechanisms for the protection against herbivores and pathogens (Lee 2013; Mahar et al. 2016). Further, unveiling the ecological roles of metal hyperaccumulation in plants may help us to clearly understand the mechanism of phytoremediation.

The selection of hyperaccumulating plants for phytoremediation purpose is a tedious task. Hence, indigenous plants are often recommended due to their high metal removal efficiency, less competition among them under local environmental conditions, and the lesser threat of becoming invasive (Ghosh and Singh 2005; Mukhopadhyay and Maiti 2010; Chandra et al. 2015). The ideal plants for phytoremediation should possess a series of characteristics: (a) ability to hyperaccumulate HMs preferably in the aboveground parts; (b) tolerance to high pH, salt, and accumulated toxic metal concentration; (c) fast growth and high biomass; (d) widespread highly branched roots; (e) easy to cultivate and harvest; and (f) resistant to diseases and pests (Vangronsveld et al. 2009; Chandra et al. 2015; Mahar et al. 2016). To date, some hyperaccumulator satisfies the abovesaid criteria to be applied for HM phytoremediation. Metal hyperaccumulating plants are often preferred over non-accumulators because they produce a high volume of metal-rich biomass and are economic to process for metal recovery and safe disposal, which have an additional eco-environmental benefit. To date, more than 500 plant species have been identified as metal hyperaccumulators. Notable representative includes the members of Brassicaceae, Asteraceae, Caryophyllaceae, Lamiaceae, Euphorbiaceae, Poaceae, etc., which have been reviewed elsewhere (Ali et al. 2013; Chandra et al. 2015; Mahar et al. 2016). Some examples of hyperaccumulating plants with their metal accumulation capacity are listed in Table 3.

6 Emerging Halophytes in Phytoremediation

Plants that tolerate and thrive in a highly saline environment with an extreme salt concentration (200 mM NaCl or more) are termed as halophytes (Liang et al. 2017). Halophytes are cultivable in soils irrigated with highly saline water/wastewater where good-quality water may not be available or limited in supply (due to high urban requirements and climate change) for crop irrigation in the arid environment (Manousaki and Kalogerakis 2011). It represents an additional eco-environmental benefit for phytoremediation. High salinity increases the mobility of HMs in soil and therefore facilitates their greater uptake and translocation from root to shoot to achieve phytoremediation (Wang et al. 2014; Liang et al. 2017). For example, the halophyte *Tamarix smyrnensis* removes 9.4, 19.7, and 38.3 μg of Cd in a solution containing 0, 0.5, and 3% NaCl, respectively (Manousaki et al. 2008).

High salt (like NaCl) concentration increases HM accumulation in halophytes either by enhancing metal mobility or by modifying root functions and alleviates

Table 3 Hyperaccumulator plants (with metal accumulation capacity) employed for phytoremediation

Plant species	Common name	Family	Polluted medium	Metal	Metal accumulation capacity (mg kg^{-1} DW)	Phytoremediation mechanism and metal accumulation compartment	Reference
Noccaea caerulescens	Blue stool	Brassicaceae	Water	Pb	1,700–2,300	Rhizofiltration (aerial parts/root)	Dinh et al. (2018)
Cannabis sativa L.	Hemp	Cannabaceae	Soil	Cd	151	Phytoextraction (aboveground plant parts)	Ahmad et al. (2016)
Alyssum markgrafii	Madwort	Brassicaceae	Soil	Ni	4,038	Phytoextraction (aboveground plant parts)	Salihaj et al. (2016)
Cannabis sativa L.	Hemp	Cannabaceae	Soil	Cu	1,530	Phytoextraction (aboveground plant parts)	Ahmad et al. (2016)
Tagetes minuta	Southern cone marigold	Asteraceae	Water	As	380.5	Phytoextraction (shoots)	Salazar and Pignata (2014)
Sorghum halepense L.	Johnson grass	Poaceae	Soil	Pb	1,406.8	Phytostabilization (reduction in rhizosphere)	Salazar and Pignata (2014)
Betula occidentalis	Water/red birch	Betulaceae	Soil	Pb	1,000	Phytoextraction (shoots)	Koptsik (2014)
Thlaspi caerulescens	Alpine pennygrass	Brassicaceae	Soil	Cd	5,000	Phytoextraction (shoots)	Koptsik (2014)
Helianthus annuus	Sunflower	Asteraceae	Soil	Pb	5,600	Phytoextraction (shoots)	Koptsik (2014)
Thlaspi caerulescens	Alpine pennygrass	Brassicaceae	Soil	Ni	16,200	Phytoextraction (shoots)	Koptsik (2014)
Brassica nigra	Black mustard	Brassicaceae	Soil	Pb	9,400	Phytoextraction (shoots)	Koptsik (2014)
Medicago sativa	Alfalfa/ lucerne	Fabaceae	Soil	Pb	43,300	Phytoextraction (shoots)	Koptsik (2014)

Pteris vittata	Chinese ladder rake	Pteridaceae	Soil	Hg	91.975	Phytoextraction (shoots)	Jianxu et al. (2012)
Corrigiola telephiifolia	Corrigiola (Spanish)	Caryophyllaceae	Soil	As	2,110	Phytoextraction (aboveground plant parts)	Garcia-Salgado et al. (2012)
Achillea millefolium	Common yarrow	Asteraceae	Soil	Hg	18.275	Phytoextraction (shoots)	Jianxu et al. (2012)
Pteris vittata	Chinese ladder rake	Pteridaceae	Water	As	20,707	Phytoextraction (shoots)	Kalve et al. (2011)
Eleocharis acicularis	Needle spikerush	Cyperaceae	Water	Zn	11,200	Phytoextraction (shoots)	Sakakibara et al. (2011)
Solanum photeinocarpum	American black nightshade	Solanaceae	Soil	Cd	158	Phytoextraction (shoots)	Zhang et al. (2011)
Eleocharis acicularis	Needle spikerush	Cyperaceae	Water	Cu	20,200	Phytoextraction (shoots)	Sakakibara et al. (2011)
Pteris vittata	Chinese ladder rake	Pteridaceae	Water	Cr	35,303	Phytoextraction (shoots)	Kalve et al. (2011)
Eleocharis acicularis	Needle spikerush	Cyperaceae	Water	Cd	239	Phytoextraction (shoots)	Sakakibara et al. (2011)
Pteris vittata	Chinese ladder rake	Pteridaceae	Water	As	20,707	Phytoextraction (shoots)	Sakakibara et al. (2011)
Eleocharis acicularis	Needle spikerush	Cyperaceae	Water	As	1,470	Phytoextraction (shoots)	Sakakibara et al. (2011)
Alyssum murale	Yellow tuft	Brassicaceae	Soil	Ni	4,730–20,100	Phytoextraction (shoots)	Bani et al. (2010)
Alyssum markgrafii	Madwort	Brassicaceae	Soil	Ni	19,100	Phytoextraction (shoots)	Bani et al. (2010)
Thlaspi caerulescens	Alpine pennygrass	Brassicaceae	Soil	Cd	3,000	Phytoextraction (shoots)	Sheoran et al. (2009)

(continued)

Table 3 (continued)

Plant species	Common name	Family	Polluted medium	Metal	Metal accumulation capacity (mg kg^{-1} DW)	Phytoremediation mechanism and metal accumulation compartment	Reference
Haumaniastrum katangense	Copper flower	Lamiaceae	Soil	Cu	8,356	Phytoextraction (shoots)	Sheoran et al. (2009)
Rorippa globosa	Globe yellowcress	Brassicaceae	Soil	Cd	>100	Phytoextraction (shoots)	Wei et al. (2008)
Azolla pinnata	Water velvet	Salviniaceae	Water	Cd	740	Rhizofiltration (bioaccumulation)	Rai (2008)
Marrubium vulgare	Common horehound	Lamiaceae	Soil	Hg	13.8	Phytoextraction (shoots)	Luis et al. (2007)

metal-induced phytotoxicity through improved management of osmotic solutes and oxidative status (Chai et al. 2013). Thus, halophytes are suitable candidates for the phytoremediation of HM-contaminated saline soils (Table 4). The role of salinity in increasing Cd bioavailability is mainly attributable to the formation of Cd-Cl complexes ($CdCl^+$ and $CdCl_2$) (Weggler-Beaton et al. 2000; Chai et al. 2013). These complexes are less strongly sorbed to soil than a free Cd^{2+} ion and hence increase Cd mobility at the soil-root interface. Moreover, these complexes also stimulate transport of Cd across the zone encompassing soil-rhizosphere apoplast-plasma membrane. Thus, increased soil-plant transfer of Cd can occur under salinity. Although the detailed interaction between salinity and HM accumulation is still not fully understood.

The halophytes applied for the phytoremediation of HM-contaminated saline soils have been recently reviewed (Oosten and Van Maggio 2015; Liang et al. 2017). The mechanism of metals and salt tolerance in halophytes include (a) osmotic adjustment through ion accumulation/compartmentalization or exclusion and biosynthesis of compatible solutes; (b) involvement of antioxidant defense system; (c) cell walls and subcellular compartmentalization; (d) metal chelation or detoxification; and (e) metal excretion and complexing ligands (Liang et al. 2017). Recently, "phytoexcretion" has been introduced as a novel phytoremediation process for salt-affected metal-contaminated sites (Liang et al. 2017). It is a type of metal detoxification strategy in halophytes, wherein toxic metals are excreted through specialized salt glands from leaf tissue onto leaf surface (Manousaki and Kalogerakis 2011; Liang et al. 2017). Thus, applying phytoextraction in conjunction with phytoexcretion using halophytes represents a promising strategy for the phytoremediation of high salt-affected metal-contaminated sites.

Halophytes such as *Atriplex halimus*, *Echinochloa stagnina*, *Spartina alterniflora*, *Zygophyllum fabago*, *Sesuvium portulacastrum*, and *Tamarix africana* have been well reported for the phytoextraction/phytostabilization of HM-contaminated saline soils (Liang et al. 2017). The use of halophyte *Atriplex nummularia* in the remediation of saline and sodic soils is also reported due to its high biomass and salt extraction capability (de Souza et al. 2014). However, the suitability of different halophytes for the phytoremediation of HM-contaminated sites is still under evaluation. The future research should be focused on the (a) mechanistic understanding of simultaneous salt and HM tolerance; (b) use of halophytes of economic importance for HM phytoremediation to gain better economic returns; (c) use of transgenics to develop engineered halophytes with high biomass and fast growth rate for effective HM phytoremediation under abiotic stress; and (d) use of biochar and other soil amendments for the improved phytoremediation of salt-affected HM-contaminated sites (Liang et al. 2017).

Table 4 Halophytes and medicinal plants reported in phytoremediation of heavy metal-contaminated soils

	Metal	Remarks	Reference
Halophytes			
Spartina alterniflora	Cu	Accumulation of metal in roots (phytostabilization)	Chai et al. (2014)
Phragmites australis	Cd	Accumulation of metal in roots (phytostabilization)	Nunes da Silva et al. (2014)
Moringa oleifera	Cd	Through leaf extraction (phytoextraction)	Howladar (2014)
Zygosaccharomyces rouxii	Cd, Zn, Cu, and Pb	Extraction and exclusion of heavy metals (phytoextraction)	Li et al. (2013)
Salicornia bigelovii	Se	Se removal up to $251.6 \pm 140.5\ \mu g/m^2 d$ (phytovolatilization)	Huang et al. (2013)
Halimione portulacoides	Zn	Accumulation of metal in tissues (phytostabilization)	Andrades-Moreno et al. (2013)
Suaeda salsa	Pb and Zn	Accumulation of metals in roots (phytostabilization)	Wu et al. (2013)
Salicornia ramosissima	Cd	Accumulation of metal in roots (phytostabilization)	Pedro et al. (2013)
Arthrocnemum macrostachyum	Cd	Accumulation of metal in roots (phytostabilization)	Redondo-Gómez et al. (2010)
Commelina communis	Cu	Accumulation of metal in roots (phytostabilization)	Wang and Zhong (2011)
Salicornia brachiata	Cd, Ni, and As	Accumulation metals in roots (phytostabilization)	Xu et al. (2010)
Sarcocornia perennis	Cd, Zn, Cu, and Co	Accumulation of metal in roots (phytostabilization)	Lefèvre et al. (2010)
Medicinal plants			
Lavandula vera L.	Cd, Pb, and Zn	Plant is a good phytoremediator for Pb, Cd, and Zn (phytoextraction)	Angelova et al. (2015)
Portulaca oleracea L.	Cr	Significant Cr accumulation in harvestable parts (phytoextraction)	Kale et al. (2015)
Portulaca oleracea L.	Pb, Ni, and Zn	Plant species is not suggested for HM's phytoremediation purpose due to inhabitation of plant roots at high metal concentration	Amer et al. (2013)
Ocimum tenuiflorum L., *Ocimum gratissimum* L., and *Ocimum basilicum* L.	As	*Ocimum* spp. can be used as a phytoremediator and at the same time as a source of essential oils under proper regulation	Siddiqui et al. (2013)
Ocimum gratissimum L.	Cd and Zn	Plant species could not be considered as hyperaccumulator due to poor metal translocation capacity from roots to shoots	Chaiyarat et al. (2011)

(continued)

Table 4 (continued)

	Metal	Remarks	Reference
Artemisia annua L.	As	*A. annua* could be considered as the top potential candidate for cultivation in As-contaminated soil	Rai et al. (2011)
Mentha arvensis L., *Mentha piperita* L., and *Mentha citrata* L.	Cr and Pb	The studied species are not hyperaccumulator for Cr and Pb	Prasad et al. (2010)
Nasturtium officinale R. Br	Ni	*N. officinale* can be utilized as a phytoremediator in moderately polluted wetlands (phytoextraction)	Duman and Ozturk (2010)

7 Medicinal and Aromatic Plants in Phytoremediation

Generally, edible crops are not suitable for the phytoremediation of HM-contaminated sites due to the risk of entering into the food chain via consumption by humans/animals and associated health hazards. The application of medicinal and aromatic plants for the phytoremediation of HM-contaminated sites could be an innovative approach. Aromatic plants are mostly nonedible and are not being consumed directly by humans or animal due to their essence. They have low metal accumulation potential as compared to edible plant hyperaccumulators but are economically important as their harvested foliage are the chief source of essential oil. The essential oil obtained from aromatic plants is free from the risk of HM accumulation from plant biomass and thus prevents entry of HMs into the food chain (Gupta et al. 2013; Lajayar et al. 2017). HMs remain in the extracted plant residues during oil extraction through the distillation process and hence limit the detectable concentration in the essential oil. The export of essential oil for selling is a major economic incentive along with phytoremediation because it is being used in soaps, detergents, insect repellents, cosmetic, perfumes, and food processing industries (Gupta et al. 2013; Lajayar et al. 2017).

Some aromatic plants, such as *Artemisia annua*, *Mentha arvensis*, *Cannabis sativa*, *Lavandula vera*, *Matricaria chamomilla*, *Mellissa officinalis*, *Ocimum gratissimum*, *Portulaca oleracea*, and *Salvia officinalis*, have been investigated for the phytoremediation of HM-contaminated soil (Lajayar et al. 2017) (Table 4). However, due to higher ability of some aromatic plants to accumulate HMs in their aerial parts, when the intention of their cultivation on contaminated soil is non-phytoremediation (i.e., edible consumption), the consumption of such aromatic plants may result in serious health hazards and thus require continuous monitoring (Lajayar et al. 2017). The phytoremediation of HM-contaminated sites using aromatic plants is a newly emerging concept and currently under research. There are very few available studies to date, and hence, additional research is required to explore their potential in phytoremediation. Furthermore, transgenic approaches could be used to enhance the metal accumulation capacity in such plants. For instance, Dhankher et al. (2002) developed transgenic *Arabidopsis thaliana* plants (medicinal herb used to cure sores in the mouth) with increased tolerance and accumulation of As for enhanced phytoremediation.

8 Molecular Mechanism of Heavy Metal Tolerance, Uptake, Translocation, and Phytoremediation

HMs are highly toxic to plants as they disturb the redox status (balance between oxidants and antioxidants) and lead to oxidative stress responsible for the physiological damage. Plants always maintain a very low concentration of free radicals to avoid any physiological damage. This balance is established by the uptake and translocation of toxic metals, their sequestration, and binding to proteins and organic ligands. Plants uptake toxic metals from polluted matrix via roots and store them either in roots or translocate to the shoots through xylem vessels, where they sequester in the vacuoles. Vacuoles help to reduce the excess metal ions from the cytosol and prevent their interactions with other metabolic processes due to low metabolic activity and thus are considered to be the ideal sites of metal sequestration (Wu et al. 2010; Ali et al. 2013; Chandra et al. 2015). Metal tolerance is a pivotal requirement for metal accumulation and phytoremediation and governed by a variety of biomolecules. Membrane transporter proteins (MTPs) such as ATPases, zinc-iron permease (ZIP), cation diffusion facilitators (CDF), copper transporters (COPTs), ATP-binding cassettes (ABC), and cation exchangers (CAXs) and heavy-metal ATPase (HMA) such as P1B-ATPase, HMA4, and HMA5 help in the uptake and transport of metals across the cell membrane and facilitate their detoxification (Ali et al. 2013; Sarwar et al. 2017). Metal-binding proteins (such as Cu chaperone ATX1-like proteins, glutathione (GSH), metallothioneins (MTs), and phytochelatins (PCs)) and organic ligands are involved in the binding, sequestration, and detoxification of toxic metals in the aboveground plant parts preferably in the cuticle, epidermis, and trichomes as these have less chance of cellular damage (Wu et al. 2010; Chandra et al. 2015; Sarwar et al. 2017). GSH protect the plants from physiological damage caused by toxic metal stress, MTs reduce metal accumulation in shoots by trapping them in roots, and PCs enhance HM tolerance, accumulation, and detoxification in plants (Goel et al. 2009; Wu et al. 2010). Thiol groups present in these biomolecules form complexes with HMs (GSH-PC-MT-HM) and thus play a crucial role in the detoxification of toxic metals in plants (Sarwar et al. 2017).

9 Exploiting Plant-Microbe Interactions for Enhanced Metal Phytoremediation

The longtime frame required for phytoremediation and physiological damage to remediating plants under toxic metal stress is a major issue. Therefore, plant-microbe interactions (PMIs) could be exploited to enhance the plant growth and phytoremediation of HM-contaminated sites.

The root-/rhizosphere-colonizing, plant growth-promoting rhizobacteria (PGPR) have been reported to enhance host plant growth in toxic metal-contaminated sites (Yuan et al. 2013; Ma et al. 2015, 2016a). PGPR produces growth hormones such as auxins (IAA: indole-3-acetic acid), cytokinins, gibberellins, and ethylene (Rajkumar et al. 2012; Ma et al. 2015). The mechanisms of plant growth promotion may vary from bacterial strain to strain and depend on various secondary metabolites produced (Ma et al. 2011; Backer et al. 2018). PGPR also produces some other beneficial compounds such as enzymes, osmolytes, biosurfactants, organic acids, metal-chelating siderophores, nitric oxide, and antibiotics (Rajkumar et al. 2012; Ma et al. 2015). These beneficial compounds reduce ethylene production via synthesis of ACC (1-aminocyclopropane-1-carboxylate) deaminase that prevents the inhibition of root elongation, lateral root growth, and root hair formation and also improve the mineral (N, P, and K) uptake in acidic soil (Babu et al. 2013; Ma et al. 2015). These compounds also suppress phytopathogens, provide tolerance to abiotic stress, and help in associated nitrogen fixation (Rajkumar et al. 2012; Babu et al. 2013; Ma et al. 2015). Hence, PGPR are applied in sustainable agriculture development. Besides these, PGPR can lower the metal toxicity to remediating plants through biosorption/bioaccumulation as bacterial cells have an extremely high ratio of surface area to volume (Ma et al. 2016b; Li et al. 2018). PGPR could adsorb high metal concentration by either a metabolism-independent passive or metabolism-dependent active processes. Hence, using PGPR in environmental bioremediation could be a useful strategy for plant survival in the stressed environment. PGPR reported for the enhanced HM phytoremediation with associated benefits have been reviewed in the past (Ma et al. 2011; Rajkumar et al. 2012; Ullah et al. 2015). Some updated examples from recent studies are summarized in Table 5.

Endophytes are the microbes (bacteria/fungi) that reside in the inner tissues of plants without causing harm to host. They also help in plant growth promotion and development under biotic- or abiotic-stressed environment and exert many beneficial effects than rhizobacteria (Luo et al. 2011; Ma et al. 2011, 2015). They are able to tolerate high metal concentration and hence lower phytotoxicity to remediating plants as well as help in growth promotion enhancing through biocontrol mechanism and induced systemic resistance against phytopathogens (Ma et al. 2011, 2015). They produce phytohormones, organic acids, siderophores, biosurfactants, enzymes, and growth regulators that help in water and nutrient (P, N, and K) uptake, osmolyte accumulation, osmotic adjustment, stomatal regulation, and associated nitrogen fixation as additional benefits to host plants (Ma et al. 2011, 2016b). Thus, inoculating plants with endophytes could be an excellent strategy to enhance the

Table 5 Recent studies on microbe-assisted phytoremediation of heavy metal-contaminated soils

Bacterial strain(s)	Host plant	Heavy metal	Metal conc. in control plant	Metal conc. in inoculated plant	Medium	Beneficial effects	Reference
Enterobacter sp. LC1, LC4, and LC6, *Kocuria* sp. LC2 and LC5 and *Kosakonia* sp. LC7	*Solanum nigrum*	As	152.3 mg kg^{-1}	2,000 mg kg^{-1}	Soil	IAA and P-solubilization	Mukherjee et al. (2018)
Pseudomonas libanensis and *Pseudomonas reactans*	*Brassica oxyrrhina*	Cu, Zn	Cu 93 mg kg^{-1}, Zn 835 mg kg^{-1}	Cu 219 mg kg^{-1}, Zn 1,254 mg kg^{-1}	Soil	IAA, ACC deaminase, siderophores	Ma et al. (2016a)
Rhizobium leguminosarum	*Brassica juncea*	Zn	Aboveground 100 mg kg^{-1}, belowground 200 mg kg^{-1}	Aboveground 1,200 mg kg^{-1}, belowground 400 mg kg^{-1}	Soil	Metal chelation	Adediran et al. (2015)
Bacillus pumilus E2S2 and *Bacillus* sp. E1S2	*Sedum plumbizincicola*	Cd	125 mg kg^{-1}	180 mg kg^{-1} (E2S2), 165 mg kg^{-1} (E1S2)	Soil	IAA, ACC deaminase, siderophores, P-solubilization	Ma et al. (2015)
Pseudomonas sp. LK9	*Solanum nigrum*	Cd	230 mg kg^{-1}	292 mg kg^{-1}	Soil	Biosurfactants, siderophores, organic acids	Chen et al. (2014)
Enterobacter sp. JYX7 and *Klebsiella* sp. JYX10	*Polygonum pubescens*	Cd	Root 65 mg Cd kg^{-1}, shoot 38 mg Cd kg^{-1}	Root 170 mg Cd kg^{-1} (JYX7), 155 mg Cd kg^{-1} (JYX10); shoot 50 mg Cd kg^{-1} (JYX7), 56 mg Cd kg^{-1} (JYX10)	Soil	IAA, siderophores, ACC deaminase, P-solubilization	Jing et al. (2014)
Bacillus thuringiensis GDB-1	*Alnus firma*	As	Root 1,911 mg kg^{-1}, shoot 762 mg kg^{-1}	Root 2,162 mg kg^{-1}, shoot 845 mg kg^{-1}	Mine tailing waste	ACC deaminase, IAA, siderophores, P-solubilization	Babu et al. (2013)
Rahnella sp. JN6	*Brassica napus*	Pb	Shoot 22 mg kg^{-1}, root 77 mg kg^{-1}	Shoot 32 mg kg^{-1}, root 113 mg kg^{-1}	Soil	IAA, ACC deaminase, siderophores, P-solubilization	He et al. (2013)

Staphylococcus arlettae NBRIEA G-6	*B. juncea*	As	Root 11 mg kg^{-1}, shoot 2 mg kg^{-1}	Root 12 mg kg^{-1}, shoot 70 mg kg^{-1}	Soil	IAA, siderophores, ACC deaminase	Srivastava et al. (2013)
Rahnella sp.	*Amaranthus hypochondriacu*, *A. mangostanus*, and *S. nigrum*	Cd	55 μg pot^{-1}	130 μg pot^{-1}, 117 μg pot^{-1}, 110 μg pot^{-1}	Soil	IAA, siderophores, ACC deaminase, P-solubilization	Yuan et al. (2013)
Paenibacillus macerans NBRFT5, *Bacillus endophyticus* NBRFT4, and *Bacillus pumilus* NBRFT9	*Brassica juncea*	Ni	300 μg $plant^{-1}$	Combined treatment: 950 μg $plant^{-1}$	Mix. of fly ash and press mud	Siderophores, organic acids, protons, and other nonspecified enzymes	Tiwari et al. (2012)
Pantoea agglomerans Jp3-3 and *Pseudomonas thivervalensis* Y1-3-9	*Brassica napus*	Cu	Shoot 8 μg pot^{-1}, root 7 μg pot^{-1}	Shoot 25 μg pot^{-1} (Jp3–3), root 20 μg pot^{-1} (Jp3–3); shoot 25 μg pot^{-1} (Y1–3-9), root 15 μg pot^{-1} (Y1–3-9)	Quartz sand	IAA, siderophores, ACC deaminase, P-solubilization	Zhang et al. (2011)
Azotobacter chroococcum and *Rhizobium leguminosarum*	*Zea mays* L.	Pb	290 μg g^{-1}	1,329 μg g^{-1}	Soil	IAA production increased and soil pH decreased	Hadi and Bano (2010)
Bacillus subtilis, *B. cereus*, *Flavobacterium* sp., and *Pseudomonas* sp.	*Orychophragmus violaceus*	Zn	Root 187 mg kg^{-1}, shoot 253 mg kg^{-1}	Root 238–377 mg kg^{-1}, shoot 290–486 mg kg^{-1}	Soil	ACC deaminase, IAA, siderophores	He et al. (2010)

phytoremediation of HM-contaminated sites. Endophytes applied to enhance HM phytoremediation with associated benefits have been recently reviewed by several researchers (Afzal et al. 2014; Ma et al. 2016b).

Arbuscular mycorrhizal fungi (AMF: colonize plant roots) have been also reported to protect their host plants against heavy metal toxicity through their mobilization from soil and thus help in phytoremediation (Marques et al. 2009; Meier et al. 2012; Khan et al. 2014). The possible mechanisms by which AMF protect their host plants through metal mobilization from soil include (a) immobilization by chelation; (b) binding of metals to biopolymers in the cell wall; (c) superficial immobilization in the plasmatic membrane once metals cross the cell wall; (d) membrane transportation that mobilizes metals from the soil to the cytosol; (e) intracellular chelation through MTs, organic acids, and amino acids; (f) export of metals from cytosol by membrane transporters; (g) sequestration of metals into vacuoles; (h) transportation of metals by means of fungal hyphae; (i) storage of metals in fungal spores; and (j) exportation by the fungus, and access into the plant cells, involving both active and passive transportation into the mycorrhizae (Meier et al. 2012; Cabral et al. 2015). They confer resistance against drought, high salt, and toxic metal concentration and improve nutrient supply and soil physical properties (Khan et al. 2014). The exact mechanism of plant protection is still not fully understood, and further research is required to explore their role in the phytoremediation. In addition, isolating and characterizing suitable plant-associated beneficial microbes is a time-consuming process. It also requires the analysis of more than thousands of isolates, and thus, identification of specific biomarkers may help to select the effective plant-microbe associationships for microbe-assisted phytoremediation (Rajkumar et al. 2012). Further, to ameliorate metal toxicity, plant growth promotion, and metal sequestration, extensive research efforts are also required to explore novel microbial diversity, their distribution, as well as functions in the autochthonous and allochthonous soil habitats for microbe-assisted phytoremediation of HM-contaminated sites.

10 Molecular Approaches for Enhanced Phytoremediation of Heavy Metals

Transgenic approaches are decisive in genetic manipulation of low biomass and slow-growing hyperaccumulating plants to enhance the phytoremediation of HM-contaminated sites. The main objective is to introduce genes (from organisms such as plants, bacteria, and mammals) that confer plants the ability to resist, tolerate, and hyperaccumulate toxic metals from contaminated sites under changeable environment with increased biomass and metal storage capacity (Goel et al. 2009; Marques et al. 2009; Mukhopadhyay and Maiti 2010). The general approach behind the transgenic approaches is to overexpress or knock down genes that encode for metal-binding PCs, metal-chelating MTs, and MTPs, which are crucial in toxic metal detoxification and thus phytoremediation (Goel et al. 2009; Mukhopadhyay and Maiti 2010; Dhankher et al. 2011; Sarwar et al. 2017).

Chloroplast engineering is an innovative approach that allows the transfer of the entire operons from bacteria into plants for the overexpression of enzymes responsible for HM phytoremediation (Dhankher et al. 2011; Goel et al. 2009). Identification of novel genes and their transfer from natural hyperaccumulators and microbes into fast-growing metal hyperaccumulators may also create new opportunities for enhancing phytoremediation efficiency. To date, transgenic plants have not yet been applied for field-scale application, and genes escaping from transgenic plants to wild relatives are very rare. Genetically engineered plants (GEPs) applied for the enhanced metal phytoremediation in the laboratory have been reviewed (Kotrba et al. 2009; Goel et al. 2009; Vangronsveld et al. 2009; Chandra et al. 2015). Some updated examples from the recent studies are summarized in Table 6. However, there are several environmental risks associated with transgenic plants applied for metal phytoremediation at the field that should also be considered. Environmental risks may include (a) exposure of toxic metals to wildlife and humans to their more bioavailable forms; (b) uncontrolled spread of transgenic plants due to interbreeding with their wild relatives or superior fitness because of weedy nature in environment (genetic pollution); (c) transformation of natural flora through cross-pollination; and (d) risk of invasion of free plants and potential loss of diversity (Kotrba et al. 2009; Marques et al. 2009; Wu et al. 2010). Hence, risk assessment should be performed before applying GEPs at the field scale.

The engineering of symbiotic plant-associated microbes (i.e., PGPRs and endophytes) could also be a promising phytobacterial technology to enhance the tolerance to high metal concentrations and detoxification (Weyens et al. 2013; Huang et al. 2016). It involves the introduction of one or more genes of interest that code for enzymes responsible for the enhanced remediation, stress tolerance, metal chelators, uptake regulators, transporters, and homeostasis (Ullah et al. 2015). Some updated examples from the recent studies on the genetically engineered bacteria applied for enhanced metal phytoremediation are listed in Table 7. However, engineering plant-associated microbes is of limited scope as these mainly concentrate around the roots, show limited distribution outside rhizosphere, and depend chiefly on the host plants (Wu et al. 2006). Hence, their controlled use outside rhizosphere is not an easy task.

Recent advances in "omics" technologies (such as proteomics, metabolomics, genomics, metagenomics, and transcriptomics) also offer greater opportunities to identify traits that maximize the benefits of phytoremediation through manipulating tolerance, accumulation, and pollutant degradation/detoxification potential of plants and microbes. Thus, the insertion and overexpression of genes and metal-binding proteins and their exploitation to increase metal-binding capacity and tolerance or accumulation of toxic metals in bacteria and plants could be an excellent strategy for the enhanced phytoremediation.

Table 6 Recent studies on transgenic plants for enhanced phytoremediation of heavy metals

Transgene[a]	Origin	Target plant	Effect(s)/enhanced trait	Reference
OsMT-3a	*Oryza sativa*	*Oryza sativa*	Enhanced tolerance to Cd	Mekawy et al. (2018)
VsCCoAOMT	*Vicia sativa*	*Arabidopsis thaliana*	Enhanced tolerance and accumulation of Cd	Xia et al. (2018)
AtACR2	*Arabidopsis thaliana* L.	*Nicotiana tabacum* var. Sumsun	Plants well survived on culture medium containing 200 μM arsenate, but control plants were barely survived; upon exposure to 100 μM arsenate for 35 days treatment, As content in shoots was significantly lower (28 μg/g d wt.) than control (40 μg/g d wt.), whereas in root, it was significantly higher (2,400 μg/g d. wt.) than control (2,100 μg/g d. wt.)	Nahar et al. (2017)
SaHMA3	*Sedum alfredii*	*Nicotiana tabacum*	Cd tolerance and accumulation	Zhang et al. (2016)
StGCS-GS	*Streptococcus thermophilus*	*Beta vulgaris* L.	Enhanced tolerance and accumulation of different concentrations of Cd, Zn, and Cu in shoot with increased biomass, root length, and relative growth compared with wild-type plants	Liu et al. (2015)
SaMT2	*Sedum alfredii*	*Saccharomyces cerevisiae/ Nicotiana tabacum*	Enhanced tolerance and accumulation of Cd	Zhang et al. (2014)
MerE	*E.coli* XL1-Blue	*Arabidopsis thaliana* L.	Enhanced tolerance and accumulation of methylmercury and mercuric ions in plants as compared to control	Sone et al. (2013)
CYP2E1 and *GST*	*Homo sapiens*	Alfalfa (*Medicago sativa* L.)	Accumulations of Hg and TCE were increased (3.0–4.2-fold for Hg, 2.1–4.0-fold for TCE) from liquid MS medium with increased resistance up to 50/500 μM of Hg/TCE (trichloroethylene) complex than control	Zhan et al. (2013)
ScYCF1	*Saccharomyces cerevisiae*	*Populus alba* X *P. tremula* var. *glandulosa*, BH1 *poplar clone*	Enhanced growth, reduced toxicity symptoms, and increased Cd content in aerial tissues from hydroponic solution than control; increased Cd, Zn, and Pb content in root than control from mine tailing soil	Shim et al. (2013)
EhMT1	*Elsholtzia haichowensis*	*Nicotiana tabacum*	Enhanced tolerance and accumulation of Cd	Xie et al. (2012)

YCF1	*Saccharomyces cerevisiae*	*Brassica juncea* L.	Seedlings showed 1.3–1.6-fold tolerance to Cd stress and 1.2–1.4-fold tolerance to Pb stress as compared to wild-type (WT) plants (per gram fresh weight) with increased accumulation of metals (1.5–2-fold higher Cd(II) and Pb(II) levels than those of WT)	Bhuiyan et al. (2011)
tcu1	*Neurospora crassa*	*Nicotiana tabacum* L.	Higher acquisition of copper (up to 3.1 times) in transgenic plants grown in hydroponic medium as compared with control plants with more Cu accumulation when grown in soil; however, supplementation of other divalent cations such as Cd^{2+} and Zn^{2+} did not alter uptake of Cu by transgenic plants	Singh et al. (2011)
tzn1	*Neurospora crassa*	*Nicotiana tabacum* L.	Enhanced accumulation of Zn (up to 11-fold higher) was reported as compared to control; however, no significant difference in uptake of Cd^{2+}, Fe^{2+}, Ni^{2+}, Cu^{2+}, Mn^{2+}, and Pb^{2+} between the transgenic and control plants was observed	Dixit et al. (2010)
PsMTA1	*Pisum sativum* L.	*Populus alba* L.	Plant survived at 0.1 mM $CuCl_2$ concentration at which the growth of control was strongly affected; root growth was slightly affected due to ectopic expression of the transgene; ROS accumulated at lower extent in leaf tissues than control with greater tolerance to photooxidative stress and low DNA damage upon exposure to Cu	Balestrazzi et al. (2009)
merP	*Bacillus megaterium*	*A. thaliana* L.	Plants were capable of germination and growth on media amended with 12.5 μM Hg^{2+}, accumulating 5.35 μg Hg^{2+}/g of fresh seedling weight	Hsieh et al. (2009)
GSH1	*S. cerevisiae*	*A. thaliana* L.	Cd accumulation increased 4 times higher from media amended with 30 ppm Cd^{2+}, and As accumulation increased 2.5 and 4.4 times higher from media amended with 28 ppm AsO_4^{3-} and AsO^{2-}, respectively; however, no effect on Cd^{2+}, AsO_4^{3-}, and AsO^{2-} tolerance was observed	Guo et al. (2008)

(continued)

Table 6 (continued)

Transgene[a]	Origin	Target plant	Effect(s)/enhanced trait	Reference
GSH1 and *AsPCS1*	*S. cerevisiae* and *A. sativum*	*A. thaliana* L.	Root length increased 2 times longer on media with 50 μM Cd^{2+}, 150 μM AsO_4^{3-}, or 50 μM AsO^{2-}; Cd accumulation increased (10 times higher from media with 30 ppm Cd^{2+}) and As accumulation increased (3 and 10 times higher) from media amended with 28 ppm AsO_4^{3-} and AsO^{2-}, respectively	Guo et al. (2008)
AtPCS1	*A. thaliana* L.	*B. juncea* L.	Root length increased 1.9 and 1.4 times longer, respectively, on media amended with 100 μM Cd^{2+} and 500 μM AsO_4^{3-}, respectively	Gasic and Korban (2007)

[a]Encoded gene products: *OsMT-3a* metallothionein, *VsCCoAOMT V. sativa* homologue of caffeoyl-CoA O-methyltransferase, *GST* glutathione S-transferase, *CYP2E1* cytochrome P450 2E1, *ScYCF1 Saccharomyces cerevisiae* cadmium factor 1, *SaHMA3* P 1B-type ATPase, $PsMT_{A1}$ type 2 MT able to bind Cu in planta with strong affinities for other metal ions such as Zn and Cd, *YCF1* yeast cadmium factor 1, *AtACR2 Arabidopsis thaliana* arsenic reductase 2, *SaMT2* metallothionein, *StGCS-GS Streptococcus thermophilus* γ-glutamylcysteine synthetase-glutathione synthetase, *EhMT1* metallothionein, *merE* a broad-spectrum mercury transporter, *MerP* Hg^{2+}-binding protein encoded within TnMERI1, *GSH1* yeast glutathione synthetase, *AsPCS1* phytochelatin synthase of garlic, *AtPCS1* phytochelatin synthase (PCS) of *Arabidopsis*, *tzn1* Zn transporter, *tcu1* a high affinity Cu transporter

Table 7 Recent studies on genetically engineered bacteria-assisted phytoremediation of heavy metals

Genetically modified bacteria	Modified gene expression	Host plant	Beneficial effects	Heavy metal(s)	Reference
Pseudomonas aeruginosa (Pse-w-MT)	Mouse metallothioneins (MTs)	*Pisum sativum* L.	IAA, siderophore production, and P-solubilization	Cd	Huang et al. (2016)
Pseudomonas putida KT2440	Phytochelatin synthase (PCS)	*Triticum aestivum*	Production of phytochelatins (PCs)	Cd, Hg, and Ag	Yong et al. (2014)
Burkholderia sp. HU001 and *Pseudomonas* sp. HU002	pTOM plasmid of *Burkholderia vietnamiensis* BU61	*Salix schwerinii* × *S. viminalis* cv. Tora	Phytoremediation of mixed Cd-toluene contaminations	Cd-toluene	Weyens et al. (2013)
Meshorhizobium huakuii subsp. rengei strain B3	Iron-regulated transporter 1 gene from *Arabidopsis thaliana* (*AT*IRT*1*)	*Astragalus sinicus*	Production of MTs and PCs, enhance nodule formation	Cu, Cd, Zn, and As	Ike et al. (2008)
Meshorhizobium huakuii subsp. rengei strain B3	MTL*4* and *AT*PCS		Production of MTs and PCs	Cd	Ike et al. (2007)
Pseudomonas putida 06909	Expression of metal-binding peptide (EC20)	*Helianthus annuus*	Production of metal-binding peptide	Cd	Wu et al. (2006)
Meshorhizobium huakuii subsp. rengei strain B3	PCSA*T*		Production of phytochelatins (PCs)	Cd	Sriprang et al. (2003)
Meshorhizobium huakuii subsp. rengei strain B3	Tetrameric human metallothionein (MTL*4*)	*Astragalus sinicus*	Production of metallothioneins (MTs)	Cd	Sriprang et al. (2002)
Enterobacter cloacae CAL2	EC 4.1.99.4	*Brassica napus*	IAA, ACC deaminase, siderophores, and antibiotics	As	Nie et al. (2002)

11 Energy Crops in Phytoremediation and Bioenergy Production

The treatment and safe disposal of huge quantity of metal-contaminated biomass associated with phytoextraction is an environmental concern. Nonedible and perennial energy crops can be used to maximize the benefits of phytoremediation. Energy crops have high density, biomass, and mechanization, fast growth rate, and short rotation time and are resistant to diseases and pests. The biomass of energy crops could be economically valorized for renewable energy (biogas, biofuels, and combustion for energy generation and heating) production to fulfill the global energy demands, which is one of the major challenges of the twenty-first century (Lavanya et al. 2012; Ahmad et al. 2016). An energy crop *Miscanthus x giganteus* has been predicted to supply up to 12% energy of the total energy need of the European Union (EU) (Fruhwirth and Liebhard 2004). Applying energy crops in phytoremediation represents as an additional eco-environmental benefit in controlling soil erosion, improving soil health, and providing the wildlife habitat (Simpson et al. 2009; Gomes 2012). It could also be an economic incentive for phytoremediation, particularly when energy crops are grown in barren metal-contaminated land that does not compete for food production. It will reduce the consumption of nonrenewable fossil fuels and creates wide employment opportunities for locals with the low-impact treatment of barren metal-contaminated lands (Gomes 2012). A list of energy crops used for the phytoremediation and biofuel/bioenergy production is presented in Table 8.

In addition to phytoremediation and bioenergy production, the energy crops also help in carbon sequestration and biodiversity management due to their giant structure, longer life cycle, and high above- and belowground biomass, which favors soil microfauna and shelters invertebrates and birds and flora (Blanco-Canqui 2010). They improve soil characteristics by improving soil organic matter, soil aggregation, water retention, hydraulic conductivity, macroporosity, nutrient recycling, and storage and fluxes of water, air, and heat and thus reduce water and wind erosion (Blanco-Canqui 2010). They also improve water quality by reducing the off-site transport of metal pollutants and thus reduce the risks of water pollution (Blanco-Canqui 2010). However, the utilization of energy crops in phytoremediation has some distinct disadvantages, which includes (a) scarcity of agricultural land and threat to food production; (b) diverse agroclimatic conditions for cultivation and difficulty in producing biofuels for the entire year; (c) long maturation phase especially in case of *Jatropha* that discourage small farmers for its cultivation on agricultural field; (d) air pollution and health hazards due to harmful gaseous emissions during the burning of contaminated biomass; and (e) cost involved in production, growth, transportation, and storage of biomass and its processing. The introduction of a new plant to the agricultural field for phytoremediation and bioenergy production may negatively affect the ecosystem by land use changes, biodiversity and nutrient loss, low yield, and finally food security issues (Pandey et al. 2016). Moreover, issues related to the transfer of toxic metals during the

Table 8 Energy crops for sustainable phytoremediation of heavy metal-contaminated soils and bioenergy production

Plant species	Common name	Family	Target pollutants	Bioenergy/fuel products	References
Arundo donax L.	Giant reed	Poaceae	Phytoremediation of Zn-, Cr-, and Pb-contaminated soil	Bioethanol production from biomass	Zucaro et al. (2016) and Barbosa et al. (2015)
Phragmites australis L.	Common reed	Poaceae	Phytoremediation of Co-, Ni-, Mo-, Cd-, Pb-, Cr-, Cu-, Fe-, Mn-, Zn-, and Hg-contaminated sites	Cellulosic butanol production	Cicero-Fernández et al. (2015) and Gao et al. (2014)
Cannabis sativa L.	Hemp	Cannabaceae	Phytoremediation of Cu-, Cd-, and Ni-contaminated soil	Biogas and solid biofuel production from biomass	Ahmad et al. (2016) and Prade et al. (2012)
Eucalyptus sp.	Eucalyptus	Myrtaceae	Phytoremediation of Cu-, Zn-, Cd-, and Pb-contaminated soil	Carbon sequestration and bioenergy production from biomass	Nirola et al. (2015) and Perez-Cruzado et al. (2011)
Ricinus communis L.	Castor bean	Euphorbiaceae	Phytoremediation of Cd-, Pb-, and DDT-contaminated soil	Biodiesel production from seed oil	Zhang et al. (2015b), Huang et al. (2011), and Lavanya et al. (2012)
Miscanthus sp.	Silver grass	Poaceae	Phytoremediation of As-, Cu-, Pb-, Ni-, Cd-, and Zn-contaminated water and soil	Bioethanol production from biomass	Bang et al. (2015) and Kim et al. (2012)
Jatropha curcas L.	Physic nut	Euphorbiaceae	Phytoremediation of Cd-, Pb-, and Zn-contaminated soil	Biodiesel production from seed oil	Chang et al. (2014) and Berchmans and Hirata (2008)
Panicum virgatum L.	Switch grass	Poaceae	Phytoremediation of Zn-, Cr-, and Cd-contaminated soil	Cellulosic butanol production	Gao et al. (2014) and Chen et al. (2012)
Populus sp.	Poplar	Salicaceae	Phytoremediation of Cd-, Cu-, Fe-, Pb-, and Zn-contaminated soil	Thermal and electrical energy from biomass	Baldantoni et al. (2014) and Das and Chaturvedi (2009)
Salix sp.	Willow	Salicaceae	Phytoremediation of Cd-, Cu-, Hg-, Pb-, and Zn-contaminated soil	Bioethanol production and electricity generation from biomass	Mleczek et al. (2010)
Zea mays L.	Corn	Poaceae	Phytoremediation of Cd-, Pb-, and Zn-contaminated soil	Thermal and electrical energy from biomass	Meers et al. (2010)

burning of contaminated biomass are the major constraints associated with bioenergy production (Gomes 2012). Hence, a thorough life cycle assessment (LCA) is required to fully understand the potential hazards of using contaminated biomass for bioenergy production with low environmental impacts. Further, the quality assessment of produced biofuels and evaluating the suitability of energy crops for biofuel production and phytoremediation of metal-contaminated sites are required before applying at the field scale.

Further, more volume reduction of contaminated biomass is required for its safe disposal and commercial success of phytoremediation. Several methods such as incineration, compaction, ashing, gasification, pyrolysis, direct disposal, and liquid extraction have been suggested for the postharvest treatment and biomass disposal (Ghosh and Singh 2005; Gomes 2012). Among the available methods, incineration (smelting) is the most acceptable eco-friendly and economically feasible method. It reduces the expenditure for the transportation of biomass but increases the risks of toxic metals leaching to a deeper layer of soil leading to groundwater pollution (Mahar et al. 2016). Pyrolysis can also be a promising method for biomass disposal as pyrolysis product (i.e., biochar) is a source of ore/metal concentrate that could be utilized for the separation and recovery of metals, i.e., phytomining, and thus can generate revenue (Ghosh and Singh 2005; Paz-Ferreiro et al. 2014; Dilks et al. 2016). The use of biochar as an additive to soil could also help to sequester carbon and thus lower the deleterious effects of human-induced climate change due to CO_2 emissions (Paz-Ferreiro et al. 2014). However, no single method is effective to date, and researches are underway to find the best disposal method for the metal-contaminated biomass. An integrated concept coupling phytoremediation with bioenergy production from contaminated biomass and subsequent metal recovery has been also proposed for the sustainable remediation process (Jiang et al. 2015).

12 Field Experiences

Despite proven success in the laboratory and academics, phytoremediation is still struggling to get a jump from laboratory to field for commercial success. Most of the phytoremediation researches are currently laboratory-based, where conditions are actually different from those in the field. At the field scale, phytoremediation is restricted by many factors such as low metal bioavailability, slow plant growth rate and biomass, reduced metal accumulation and tolerance, etc. and several technical difficulties, which need to be caught up (as discussed in Sect. 7). Thus, the selection of suitable plant species for the phytoremediation of HM-contaminated sites is not an easy task. In addition, due to its time-taking nature, remediation industries/companies usually lost their interest in phytoremediation technology to take it up for commercial applications. Applying transgenic plants in phytoremediation is extremely challenging due to their highly invasive nature and risk of contaminating nontarget species with their pollens but may well remediate the contaminated sites. However, strict US or western country regulations on their release for field applications, lengthy environmental impact assessment, and high cost (approx. \$100–150 million) and long time (approx. 10 years.) period are the key constraints to get a GEP to market (Beans 2017). Lack of clear-cut understanding about phytoremediation among remediation practitioners is also a key concern. Phytoremediation is being investigated at field scale worldwide. Some updated examples from recent field studies with field experiences are summarized in Table 9.

Reddy et al. (2017) conducted a field trial on the phytoremediation of HMs and polyaromatic hydrocarbon (PAH)-contaminated slag fill site (Big Marsh, Calumet region, near Chicago, IL, USA) for 3 years. According to the study, there was no significant decrease in HM concentrations in soil (no phytoextraction), but HMs were immobilized by native grasses in combination with compost amendment applied to the soil (phytostabilization). However, PAHs were well-degraded (rhizodegradation) and thus reduced the risk of contaminants to public and environment. Doni et al. (2015) remediated the polluted marine sediments at pilot scale (Port of Livorno, Central Italy) using three selected plants (*Paspalum vaginatum* Sw., *Tamarix gallica* L., and *Spartium junceum* L.) in association with compost to remove HMs (Zn, Cu, Cd, Ni, and Pb). However, Ni and Pb were the lowest translocated metals, and the process was largely considered phytostabilization and phytoextraction to a lesser extent. Khaokaew and Landrot (2014) remediated the Cd-contaminated agricultural field (Mae Sot contaminated field, Mae Sot District, Tak Province, Thailand) using Cd-hyperaccumulating plants except for *Chromolaena odorata*, within 2 months (phytoextraction) under greenhouse condition. Willscher et al. (2013) performed a field study to remediate uranium and other HMs such as Al, Ni, and Zn from a uranium mining site (Gessenwiese, Ronneburg, Eastern Thuringia, Germany) for 14 months using hyperaccumulators *Helianthus annuus*, *Triticale*, and *Brassica juncea* in association with calcareous topsoil, mycorrhiza, and bacterial culture, and harvested plant biomass was utilized for energy production. According to experience, a very low metal accumulation was reported in case of *Triticale*; *H. annuus* accumulated high Al, whereas *B. juncea* accumulated high Zn and Ni (phytoextraction). Vamerali et al. (2009) reported that phytoextraction of elevated amounts of metals or As from Torviscosa cinder waste site using woody plants *Populus* and *Salix* sp. is not feasible. This was largely due to low productivity and low levels of translocation of metals from roots and despite an apparent high mobility of Pb and Zn. Thus, the selection of suitable plants is critical for the effective phytoremediation of HM-contaminated sites.

Further, more long-term field studies are required to document time and cost data for the economic evaluation of in situ phytoremediation at the field scale. Phytoremediation (phytoextraction) is currently under investigation in the US Environmental Protection Agency (US EPA)-supported Superfund Innovative Technology Evaluation (SITE) program (http://www.epa.gov/superfund/sites) to fully understand its economic feasibility, better performance, and wider acceptance. Moreover, phytoremediation is also under investigation in the EU demonstrative projects such as Phytosudoe (http://www.phytosudoe.eu/) or Life RiverPhy (http://liferiverphy.eu/web/en/) to evaluate its success in environmental decontamination at the field scale.

Table 9 Recent field studies on phytoremediation of heavy metal-contaminated sites

Contaminated field/site	Type of field/site	Metal pollutants	Amendment applied	Plant species	Treatment time	Remediation technique	Field experiences	References
"Big Marsh," Calumet region, near Chicago, IL, USA	Slag fill site	PAHs and heavy metals	Compost	Many plant species (please see paper)	3 years	Rhizodegradation and phytostabilization	Insignificant uptake of HMs by plants, degradation or immobilization of contaminants	Reddy et al. (2017)
Port of Livorno, Central Italy	Polluted marine sediments	Cd, Zn, Cu, Pb, and Ni	Compost	*Paspalum vaginatum* Sw.; *P. vaginatum* Sw. + *Spartium junceum* L.; and *P. vaginatum* Sw. + *Tamarix gallica* L.		Phytoextraction and phytostabilization	Best uptake of Cd, Zn, and Cu by plants, whereas Ni and Pb were the lowest translocated metals	Doni et al. (2015)
Bouwer, Province of Córdoba, Argentina	Agricultural and residential area	Pb		Many plant species (please see paper)		Phytoextraction and phytostabilization	*T. minuta* and *B. pilosa* were able to accumulate more Pb than other studied plants and thus suitable phytoremediation of Pb-polluted soils	Salazar and Pignata (2014)
Mae Sot contaminated fields, Mae Sot District, Tak Province, Thailand	Agricultural field	Cd		*Gynura pseudochina*, *Chromolaena odorata*, *Conyza sumatrensis*, *Crassocephalum crepidioides*, and *Nicotiana tabacum*	2 months	Phytoextraction	All plant species, except *C. odorata*, demonstrated Cd-hyperaccumulating capacities and well treated most Cd-impacted soils at Mae Sot	Khaokaew and Landrot (2014)

"Gessenwiese," Ronneburg, Eastern Thuringia, Germany	Uranium mining site	U, Al, Ni, and Zn	Calcareous topsoil, mycorrhiza, and bacterial culture	*Helianthus annuus*, *Triticale*, and *Brassica juncea*	14 months	Phytoextraction and phytostabilization	Very low metal accumulation was reported in case of *Triticale*; *H. annuus* accumulated high Al, whereas *B. juncea* accumulated high Zn and Ni in shoots; biogas and bioethanol production from biomass	Willscher et al. (2013)
Carpiano, Milano, and Torviscosa, Udine, Italy	Agriculture land	Cd, Cr, As, Zn, Pb, Co, Mn, Ni, and Cu	Cement, lime, lime + iron sulfate, EDTA, EDDS, NTA, manure, and biochar	Many plant species (please see paper)	10 years	Phytoextraction and phytostabilization	Chelators enhanced metal accumulation in plants from soil with increased biomass	Vamerali et al. (2012)
Torviscosa, Udine, NE Italy	Industrial waste-contaminated area	As, Cu, Co, Pb, Zn, and pyrite wastes		*P. alba* L., *P. nigra* L., *P. tremula* L., and *Salix alba* L.	2 years	Phytoextraction	Insignificant accumulation of As from waste; uptake of metals and As to aboveground tissues is marginal, and there is little risk of food web contamination via herbivores	Vamerali et al. (2009)
Lechang site, Lechang City, Guangdong Province, China	Mining site	Pb, Cd, Zn, and Cu	EDTA, NH_4NO_3, and $(NH_4)_2SO_4$	*Sorghum bicolor* L.	4 months	Phytoextraction	EDTA-enhanced accumulation of Pb, whereas NH_4NO_3 and $(NH_4)_2SO_4$ enhanced accumulation of both Zn and Cd in roots of sorghum plants	Ping et al. (2009)

(continued)

Table 9 (continued)

Contaminated field/site	Type of field/site	Metal pollutants	Amendment applied	Plant species	Treatment time	Remediation technique	Field experiences	References
Torviscosa, NE Italy	Industrial waste-contaminated area	Cd, As, Zn, Pb, Co, and Cu		*Sorghum bicolor* and *Helianthus annuus*	40 days	Phytoextraction	Zn removal by *S. bicolor* and *H. annuus* reached about 2,000 g ha^{-1} and 1,000 g ha^{-1}, respectively	Marchiol et al. (2007)
Torviscosa, NE Italy	Industrial waste-contaminated area	As, Cd, Cu, Pb, and Zn		*Glycine max*, *Sorghum bicolor*, *Zea mays*, and *Helianthus annuus*	50 days	Phytoextraction	*S. bicolor* appeared more efficient than *H. annuus* in terms of metal uptake	Fellet et al. (2007)

13 Emerging Phytotechnologies

Phytotechnology is an emerging field that implements solutions to scientific and engineering problems using plants to control and minimize environmental pollution. Phytotechnologies may provide more efficient alternatives to traditional cleanup methods because of their low capital costs and maintenance requirements, high success rates, end-use value, and aesthetic nature. Some phytotechnologies associated with phytoremediation are briefly discussed below:

13.1 Phytomining

It is a plant-assisted mining and recovery of precious metals from the ash of combusted-contaminated biomass and thus can generate revenue. If it is applied to the agricultural field, it is termed as agromining (Sheoran et al. 2009; Mahar et al. 2016; Jiang et al. 2015). Bioenergy generation and less SO_x emission due to the low sulfur content of bio-ores are the additional eco-environmental benefit of phytomining as compared to conventional mining technologies (Ali et al. 2013). It may be limited by the plant's phytoextraction efficiency and the market price of metals to be processed. It is more suitable for Au, Tl, Co, and Ni due to their high price and concentration in biomass (Mahar et al. 2016). For instance, it has been commercialized for Ni because hyperaccumulator plants such as *Alyssum murale* and *Alyssum corsicum* can accumulate 400 kg Ni ha^{-1} with a production cost of \$250–500 ha^{-1} (Ali et al. 2013). Hence, it is useful for the treatment and management of Ni-contaminated sites. Another successful case study on the phytomining is the use of hyperaccumulator *Berkheya coddii* for the phytoremediation Ni-contaminated soils near an industrial plant in Rustenburg, South Africa (Antony et al. 2015). They reported a high yield of 20 t/ha and active Ni absorption 2–3% with metal accumulation in the ash (15%) make its profitable for the repeated extraction process. The profitability of Ni phytomining using *B. coddii* on Ni-rich serpentine soils (Australia) is estimated at 11,500 AU\$/ha/yield, and the profitability of Au phytomining using *B. juncea* is about 26,000 AU\$/ha/yield (Mahar et al. 2016). However, the high market price of uranium and its low concentration (100 mg/kg) in the biomass (10 t/ha) of *Atriplex* species make the phytomining unprofitable (Sheoran et al. 2009).

Phytomining can be more economically feasible if combined with bioenergy production, and sale of C-credit could be a possible benefit (Mahar et al. 2016). Applying energy crops for the phytoremediation of contaminated sites could also pave the way for economical phytomining of valuable metals. According to a study, the cultivation of energy maize in the Campine region of the Netherlands and Belgium could result in the generation of 290,000–38,000 KV of renewable energy per hectare (Meers et al. 2010). It may reduce the need for coal-powered energy and minimize the CO_2 emission up to 21 tons/ha/year. However, phytomining is not successful in the northern regions of the world because of low plant productivity under harsh weather conditions. It is mainly suitable for the treatment and restoration of the disturbed soils, mine tailing waste, and mining sites in tropical regions and has been accepted by public and commercial enterprises. Thus, phytomining could be an economic incentive for low-cost metal recovery and pollution control.

13.2 Constructed Wetlands

The use of constructed wetlands (CWs) for wastewater treatment and management has been increasingly recognized worldwide. CWs are the man-engineered systems constructed to utilize the natural processes of aquatic macrophytes with their associated microbial assemblages for wastewater treatment within a more controlled environment (Stottmeister et al. 2003; Khan et al. 2009). CWs are mainly vegetated with different wetland plants having high biomass, fast growth rate, and metal accumulation capacity such as *Phragmites australis*, *Typha latifolia*, *Canna indica*, *Stenotaphrum secundatum*, *Scirpus americanus*, *Scirpus acutus*, *Iris pseudacorus*, etc., for the treatment of metal-rich wastewaters (Bharagava et al. 2017c). CWs have been proved to be successful in the removal of a variety of organic and inorganic pollutants such as metals, nutrients, and a wide range of micro-pollutants, such as pharmaceutical and personal care products, and also fecal indicator bacteria and pathogens (Zhang et al. 2015a). The pollutant's removal efficiency of CWs mainly depends on wastewater treatment rate, organic loading rate, hydrologic regime, hydraulic retention time, operational mode, and vegetation type (Zhang et al. 2015a). The application of CWs in pollutant's removal from wastewaters has been recently reviewed by many workers (Zhang et al. 2014, 2015a; Bharagava et al. 2017c).

CWs may provide many ecological and economic benefits such as requiring low capital investment for construction, low electricity for operation, and less maintenance and providing wildlife habitat and human recreational opportunities and a reuse and recycling option for the wastewater treatment facility. CWs are more favored in developing countries due to easily available and less costly land and tropical environment, which help to flourish the microbial communities responsible for the degradation/detoxification of organic and inorganic contaminants in wastewaters (Zhang et al. 2015a). Thus, the increasing use of CWs can successfully treat/detoxify HM-rich wastewaters and solve various water quality issues in the world. Integrating CW plants with a microbial fuel cell (MFC) for wastewater treatment and electricity generation could be an innovative approach for the improved degradation of pollutants. According to a recent study, a maximum current density of 55 mA m^{-2} could be achieved during the removal of hexavalent chromium (Cr^{6+}) from solution with greater removal efficiency (up to 90%) in an integrated plant-microbial fuel cell (PMFC) system planted with a wetland plant *Lolium perenne* (Habibul et al. 2016). Moreover, CWs may have great potential for bioenergy production and carbon sequestration, if planted with energy crops. According to a recent study, the incineration of harvested biomass (16,737 kg with C content: 6,185 kg) of *Ludwigia* sp. and *Typha* sp. recovered from a subtropical constructed wetland could produce 11,846 kWh energy for 1 month (Wang et al. 2011). Currently, researches are underway to expand the scope and efficacy of CWs for treatment of metal-contaminated wastewaters. However, future research should be focused on the following points: (a) understanding of microbiological dynamics and correlation of biological and non-biological processes in CWs; (b) knowledge of the dynamics of nutrient cycle to understand the fundamental processes of greenhouse gas emission in CWs; and (c) understanding of microbial community and plant-microbe interactions to know the underlying mechanism of pollutant removal in CWs (Carvalho et al. 2017).

14 Challenges and Future Research Prospects

Phytoremediation has untapped potential to apply in developing countries because of its low-cost and solar-driven nature. However, its field applicability gets limited by low metal phytoavailability, biomass, slow growth rate, and unavailability of target metal hyperaccumulators. Such drawbacks together span a long time period for phytoremediation to achieve the desirable remediation goal. Therefore, phytoremediation requires a high cost for treatment, safety, and liability of risks involved because long time also adds additional cost while evaluating economic feasibility at the field (Maestri and Marmiroli 2011; Conesa et al. 2012; Mahar et al. 2016). The application of synthetic chelators to achieve higher metal accumulation by plants can also be costly and may lead to undesirable environmental consequences such as disruption of physicochemical properties of soils by dissolving mineral components, toxicity to soil microorganisms and plants, and unacceptable leaching to groundwater. Hence, the chelate-assisted phytoremediation of HM-contaminated sites seems to be impractical. For instance, EDTA costs \$30,000 ha^{-1} to accumulate 10 g Pb kg^{-1} dry weight in shoots, and more readily degradable chemicals are also sometimes very costly (Chaney et al. 2007; Stephenson and Black 2014). The field applications of metal mobilizing amendments such as EDTA have been also banned in many European countries (Vangronsveld et al. 2009). The growth of plants during phytoremediation leads to changes in soil (e.g., changes in pH, increases in organic acids) that can make metals more bioavailable to the food chain before they can be remediated (Gerhardt et al. 2016), thus posing environmental risks that can negative some of the positive effects of phytoremediation. The accumulation of inorganic contaminants *in planta* can lead to their re-release, or release of their toxic forms, into the soil via leaf litter. For instance, litter from *Populus alba* that accumulate high levels of Cd and Zn in leaves is deemed to be problematic in the environment (Hu et al. 2013). Moreover, metals that are phytostabilized are not removed from the soil, and ever-changing soil conditions may lead to contaminant re-release in the environment, and hence, long-term monitoring of the site may be required to avoid land use changes in the future (Gerhardt et al. 2016). Phytoremediation efficiency may also be improved by applying genetically modified plants, but such application may also increase the cost of phytoremediation because contaminated sites require greater maintenance, monitoring, and disposal of biomass due to strict existing government regulations on their use in the field (Maestri and Marmiroli 2011; Stephenson and Black 2014). The lack of funding from public and private sector agencies for supporting further phytoremediation research is also a major challenge.

Despite several challenges, phytoremediation remains a promising technology with lots of prospects for future research. In recent years, increasing use of phytotechnologies advances the means of phytoremediation by integrating ecological engineering using plants. Phytotechnologies like CWs may provide an eco-technological solution for pollutant removal from wastewaters, and income can be generated, if vegetated with local plants such as common reeds, *Phragmites australis*, and elephant grass, *Pennisetum purpureum*, which are being used to produce goods (Stephenson and Black 2014). The development of CWs for wastewater treatment in developing countries is not well reported as in the case of European and American countries. But, Putrajaya wetland in Putrajaya city of Malaysia represents an excellent example of the commercial success of CWs in developing countries (Mahmood et al. 2013; Stephenson and Black 2014). Fluctuating redox conditions could be helpful in microbial precipitation of iron and sulfate and enhanced degradation of organic pollutants in the rhizosphere of aquatic macrophytes. Thus, it represents an innovative strategy to overcome the limitations of biotechnology and synthetic chelators (Stottmeister et al. 2003; Stephenson and Black 2014).

Exploiting transpiration as a phytotechnology in phytoremediation could be an excellent strategy to control migration of subsurface water along with metal pollutants and is termed as hydraulic control or phytohydraulics (Robinson et al. 2003; Stephenson and Black 2014). Evapotranspiration caps (ETCs) are of great significance in pollution control and management. ETCs are the vegetation cover created over the polluted matrix to prevent the migration of contaminated water from it. These do not destroy or remove contaminants but prevent the spreading of contaminants and thus protect people and wildlife (https://www.epa.gov/remedytech/citizens-guide-evapotranspiration-covers). ETCs are inexpensive as compared to typical prescriptive covers and can save up to \$120,000–180,000 ha^{-1} area. However, the testing, modeling, and monitoring of these systems may increase the final cost (Stephenson and Black 2014). "Ecolotree Cap" of the USA represents the first successful example of commercial ETC composed of fast-growing and deep-rooted trees that cover landfills and contaminated soils (Ecolotree 2013). Combining dendroremediation (tree as phytoremediator) with phytostabilization could be an excellent strategy for phytomanagement of contaminated soil. It may also increase the tangible value of land by increasing the provision of wood, feed products, and bioenergy production (Robinson et al. 2003; Conesa et al. 2012). Endophytic phytoaugmentation could also be a promising phytotechnology for treating contaminated wastewaters. In this technology, the remediating plants are augmented with potential endophytes for the balanced plant-microbe interactions and enhanced remediation efficiency. However, the slow action, season-dependent effectiveness, and lack of suitable monitoring methods are the key associated constraints that need to be addressed in the future for a successful application (Redfern and Gunsch 2016).

Combining phytoremediation with electrokinetic remediation (using low-voltage direct electric current to remove organic and inorganic pollutants from contaminated medium) could be an excellent strategy to enhance the metal mobility in contaminated soil and facilitate their plant uptake and thus phytoremediation. For instance, Mao et al. (2016) evaluated the feasibility of electrokinetic remediation coupled with phytoremediation to remove Pb, As, and Cs from contaminated paddy soil. Results revealed that the solubility and bioavailability of Cs and As were significantly increased by the electrokinetic field (EKF) and thereby lower the pH of contaminated soil. Furthermore, they observed that EKF significantly enhanced the bioaccumulation of As and Cs in plant roots and shoots and thus enhanced phytoremediation efficiency. The optimization of electrical parameters such as electrical field intensity, current application mode, the distance between the electrodes, and stimulation period and their effect on the mobility and bioavailability of HMs are the associated key challenges (Mao et al. 2016). However, the application of electrokinetic phytoremediation for the mixed contaminants (organic and inorganic) is also not reported so far (Cameselle et al. 2013). Linking phytomining with conventional mining technologies should also be used for the selective recovery of precious metals from contaminated soils to achieve commercial success (Robinson et al. 2003; Sheoran et al. 2009; Stephenson and Black 2014). Thus, phytotechnologies may provide a way for the sustainable management of HM-contaminated sites.

15 Innovative Ideas and Suggestions for Successful Phytoremediation Practices and Applications at HM-Contaminated Sites

In this section, the innovative ideas and constructive suggestions for the greater acceptance of phytoremediation to be effectively applied for the treatment and management of HM-contaminated sites are provided as below (Gerhardt et al. 2016):

1. More long-term field studies should be supported by public and private funding agencies and published in the refereed journals indicating whether or not the remediation was sufficient to meet regulatory compliance at a contaminated site. It will provide some assurance for site managers to choose appropriate remediation options at a given site.
2. Authors should indicate cost data and estimate of any valorization of biomass in published field studies that allow cost comparison with conventional remediation methods and provide a certainty for site managers to choose appropriate remediation options at a given site.

3. The low-cost use, wide applications, and social benefits of phytoremediation should be publicized in social media (via YouTube, Facebook, Twitter, and blogs) in addition to scientific journals to educate and engage stakeholders.
4. Formal (via conference presentations) and informal (via lunch) meetings with industry persons and stakeholders should be made to discuss its potential and deployment in managing contaminated sites.
5. The terminology relevant to phytoremediation should be standardized and simplified to make its commercial image/brand, engage non-remediation practitioners, and develop products, services, and technologies.
6. The cost of landfilling should be increased as it is a less desirable option for the management of contaminated soils. It will inflate the cost and encourage the private sector to explore phytoremediation and other options.
7. High biomass plants can be used in conjunction with microorganisms (rhizobacteria, ectomycorrhiza, and endophytes) to shorten the time frame and cost required for phytoremediation, allowing rapid turnover of the land for reuse.
8. Decision support tools (numerous models and decision trees) can be employed to assess the applicability (cost-effectiveness and the likelihood of meeting regulatory criteria) of phytoremediation for a given site.
9. Avoid overselling phytoremediation technology in terms of deployment at contaminated sites because it is not a quick remediation and may not produce satisfactory results at the initial stage. It requires optimization of biology and improvement in soil quality and then can be successfully deployed at the contaminated site.

16 Conclusions

1. HM pollution in the environment and associated toxicity in living beings is of serious eco-environmental concern.
2. Phytoremediation is an emerging eco-sustainable and clean-green solution for the eco-restoration of HM-contaminated sites.
3. Ecological roles of metal hyperaccumulation in plants are still unclear and required to clearly understand the phytoremediation mechanism.
4. Selecting target plants among known metal hyperaccumulators and exploring new plants for successful phytoremediation is an ongoing challenge.
5. Investigations on synthetic chelator-induced toxicity in remediating plants and fate, dynamics, and decomposition of metal-chelator complexes in the rhizosphere are required. The use of cheaper, eco-friendly, nontoxic, and degradable organic soil amendments is recommended to minimize environmental pollution, reduce toxicity to remediating plants, and enhance phytoremediation efficiency.
6. Inoculation of plants with associated microbes (such as PGPRs and endophytes) exhibiting multiple traits could be an excellent strategy to enhance metal detoxification in the rhizosphere. A clear-cut understanding of plant-microbe-metal-soil interactions is crucial for microbe-assisted phytoremediation of HM-contaminated soils.

7. Genetic engineering of metal-accumulating plants and associated microbes with required traits could be a very useful strategy for the enhanced phytoremediation, but associated risks should also be considered before field application.
8. Linking energy crops with phytoremediation could be an economic incentive for biofuel/energy production and metal recovery with many eco-environmental benefits; however, quality assessment (free from toxic metals) of produced biofuels is strictly advised.
9. Exploiting stress-tolerant medicinal and aromatic plants for metal phytoremediation could be an economically viable approach; however, some are edible plants, and thus, their use under strict monitoring is recommended for public health protection.
10. Search for new and reevaluating existing methods for the posttreatment of contaminated biomass (processing, volume reduction, and safe disposal) is suggested to gain better economic returns.
11. Integrating phytoremediation with phytotechnologies or other remediation methods will be helpful in the commercial success of phytoremediation.
12. Input from different fields of science, engineering, and technology is required to support the multidisciplinary research in phytoremediation.
13. More long-term field trials are required to document time and cost data to provide recommendations and convince regulators, decision-makers, and the general public about the low-cost applicability of phytoremediation to contaminated sites and for better acceptance in remediation industries.

Conclusively, phytoremediation is an eco-technological solution for HM pollution control and management and thus promotes sustainable development. However, in the future, phytomining (i.e., phytoextraction and recovery of precious metals (Au and Ni)) may successfully eliminate or reduce the need of conventional mining at large scale and thus can generate revenue and wide employment opportunities. Further, performance evaluation, complete utilization of by-products, and overall economic feasibility would always be the key criteria for global acceptance of phytoremediation technologies in waste management industries.

17 Summary

This paper highlights environmental pollution and toxicity profile of HMs and attempts to examine the feasibility of phytoremediation on HM-contaminated sites; explore current scientific progress, field studies, and sustainability issues; and revise the world over trends in phytoremediation research. It can be elucidated that the selection of suitable plants among known metal hyperaccumulators is a key to successful phytoremediation of HM-contaminated sites. Several important factors such as plant biomass, growth rate, tolerance to metal toxicity, the bioavailability of HMs, and varying climate conditions should be considered in the field as these together can make the implementation of phytoremediation difficult. However, search for new plants with required superior traits for the phytoremediation of HM-contaminated sites is an ongoing challenge. Phytoremediation still needs to overcome certain barriers before it can be accepted by the remediation industries. These barriers include the risk assessment of using transgenic plants for phytoremediation at field scale; definite understanding of plant-microbe-metal-soil interactions for the enhanced phytoremediation of HM-contaminated soils; strategic linkage of energy crops with phytoremediation for metal-free biofuel/energy production and metal recovery for environmental sustainability; integrating phytoremediation with other remediation technologies for greater remediation efficiency; and need of more long-term field trials to make it a commercially viable and acceptable technology. Furthermore, a collaborative framework involving experts from different spheres of research fields such as plant science, biogeochemistry, agriculture and environmental science, environmental engineering, environmental microbiology, and genetic engineering is required for the commercial success of phytoremediation.

Acknowledgments The financial support as "Major Research Projects" (Grant No.: EEQ/2017/000407) from the "Science and Engineering Research Board" (SERB), Department of Science and Technology (DST), Government of India (GOI), New Delhi, India, and University Grants Commission (UGC) Fellowship received by Mr. Gaurav Saxena for doctoral studies is duly acknowledged. Authors are also extremely thankful to the anonymous reviewers for valuable comments/suggestions for improvement in the review article.

Declarations of Interest Authors have no conflict of interest.

References

Adediran GA, Ngwenya BT, Mosselmans JFW, Heal KV, Harvie BA (2015) Mechanism behind bacteria induced plant growth promotion and Zn accumulation in *Brassica juncea*. J Hazard Mater 283:490–499

Afzal M, Khan QM, Sessitsch A (2014) Endophytic bacteria: prospects and applications for the phytoremediation of organic pollutants. Chemosphere 117:232–242

Ahmad R, Tehsin Z, Malik ST, Asad SA, Shahzad M, Bilal M, Shah MM, Khan SA (2016) Phytoremediation potential of hemp (*Cannabis sativa* L.): identification and characterization of heavy metals responsive genes. Clean Soil Air Water 44:195–201

Ali H, Khan E (2017) Environmental chemistry in the twenty-first century. Environ Chem Lett 15(2):329–346

Ali H, Khan E (2018a) What are heavy metals? Long-standing controversy over the scientific use of the term 'heavy metals' - proposal of a comprehensive definition. Toxicol Environ Chem 100:6–19

Ali H, Khan E (2018b) Trophic transfer, bioaccumulation and biomagnification of non-essential hazardous heavy metals and metalloids in food chains/webs: concepts and implications for wildlife and human health. Hum Ecol Risk Assess. https://doi.org/10.1080/10807039.2018.1469398

Ali H, Khan E, Sajad MA (2013) Phytoremediation of heavy metals-concepts and applications. Chemosphere 91:869–881

Amer N, Chami ZA, Bitar LA, Mondelli D, Dumontet S (2013) Evaluation of *Atriplex halimus*, *Medicago lupulina* and *Portulaca oleracea* for phytoremediation of Ni, Pb, and Zn. Int J Phytoremediation 15(5):498–512

Andrades-Moreno L, Cambrollé J, Figueroa ME, Mateos-Naranjo E (2013) Growth and survival 739 of *Halimione portulacoides* stem cuttings in heavy metal contaminated soils. Mar Pollut Bull 75(1–2):28–32

Angelova VR, Grekov DF, Kisyov VK, Ivanov KI (2015) Potential of lavender (*Lavandula vera* L.) for phytoremediation of soils contaminated with heavy metals. Int J Biol Biomol Agric Food Biotechnol Eng 9:479–486

Antoniadis V, Levizou E, Shaheen SM, Ok YS, Sebastian A, Baum C, Prasad MNV, Wenzel WW, Rinklebe J (2017) Trace elements in the soil-plant interface: phytoavailability, translocation, and phytoremediation - a review. Earth Sci Rev 171:621–645

Antony VDE, Alan JM, Baker RDR, Rufus LC, Christopher WN, Anderson JA, Meech PD, Erskine MO, Simonnot JV, Vaughan J, Louis M, Guillaume E, Bruno F, Qiu R, David RM (2015) Agromining: farming for metals in the future? Environ Sci Technol 49:4773–4780

Babu AG, Kim JD, Oh BT (2013) Enhancement of heavy metal phytoremediation by *Alnus firma* with endophytic *Bacillus thuringiensis* GDB-1. J Hazard Mater 250:477–483

Backer R, Rokem JS, Ilangumaran G, Lamont J, Praslickova D, Ricci E, Subramanian S, Smith DL (2018) Plant growth-promoting rhizobacteria: context, mechanisms of action, and roadmap to commercialization of biostimulants for sustainable agriculture. Front Plant Sci 9:1473

Baldantoni D, Cicatelli A, Bellino A, Castiglione S (2014) Different behaviours in phyto-remediation capacity of two heavy metal tolerant poplar clones in relation to iron and other trace elements. J Environ Manag 146:94–99

Balestrazzi A, Bonadei M, Quattrini E, Carbonera D (2009) Occurrence of multiple metal-resistance in bacterial isolates associated with transgenic white poplars (*Populus alba* L.). Ann Microbiol 59:17–23

Bang J, Kamala-Kannan S, Lee KJ et al (2015) Phytoremediation of heavy metals in contaminated water and soil using *Miscanthus* sp. Goedae-Uksae 1. Int J Phytoremediation 17:515–520

Bani A, Pavlova D, Echevarria G, Mullaj A, Reeves RD, Morel JL, Sulce S (2010) Nickel hyperaccumulation by the species of *Alyssum and Thlaspi* (Brassicaceae) from the ultramafic soils of the Balkans. Bot Serb 34:3–14

Barbosa B, Boléo S, Sidella S, Costa J, Duarte MP, Mendes B, Cosentino SL, Fernando AL (2015) Phytoremediation of heavy metal-contaminated soils using the perennial energy crops *Miscanthus* spp. and *Arundo donax* L. Bioenergy Res 8:500–1511

Beans C (2017) Phytoremediation advances in the lab but lags in the field. PNAS 114:7475–7477

Beg KR, Ali S (2008) Chemical contaminants and toxicity of Ganga river sediment from up and down stream area at Kanpur. Am J Environ Sci 4(4):362–366

Berchmans HJ, Hirata S (2008) Biodiesel production from crude *Jatropha curcas* L. seed oil with a high content of free fatty acids. Bioresour Technol 99:1716–1721

Bharagava RN, Chowdhary P, Saxena G (2017a) Bioremediation: an ecosustainable green technology: its applications and limitations. In: Bharagava RN (ed) Environmental pollutants and their bioremediation approaches, 1st edn. CRC Press, Boca Raton, pp 1–22. https://doi.org/10.1201/9781315173351-2

Bharagava RN, Saxena G, Mulla SI, Patel DK (2017b) Characterization and identification of recalcitrant organic pollutants (ROPs) in tannery wastewater and its phytotoxicity evaluation for environmental safety. Arch Environ Contam Toxicol 14:1–14. https://doi.org/10.1007/s00244-017-0490-x

Bharagava RN, Saxena G, Chowdhary P (2017c) Constructed Wetlands: an emerging phytotechnology for the degradation and detoxification of industrial wastewaters. In: Environmental pollutants and their bioremediation approaches. CRC Press, Boca Raton, pp 397–426

Bhardwaj R, Gupta A, Garg JK (2017) Evaluation of heavy metal contamination using environmetrics and indexing approach for River Yamuna, Delhi stretch, India. Water Sci 31:52–66

Bhuiyan MSU, Min SR, Jeong WJ, Sultana S, Choi KS, Song WY, Lee Y, Lim TP, Liu JR (2011) Overexpression of a yeast cadmium factor 1 (YCF1) enhances heavy metal tolerance and accumulation in *Brassica juncea*. Plant Cell Tissue Organ Cult 105:85–91

Bini C, Maleci L, Wahsha M (2017) Potentially toxic elements in serpentine soils and plants from Tuscany (Central Italy). A proxy for soil remediation. Catena 148:60–66

Blanco-Canqui H (2010) Energy crops and their implications on soil and environment. Agron J 102:403–419

Cabral L, Soares CR, Giachini AJ, Siqueira JO (2015) Arbuscular mycorrhizal fungi in phytoremediation of contaminated areas by trace elements: mechanisms and major benefits of their applications. World J Microbiol Biotechnol 31(11):1655–1664

Cameselle C, Chirakkara RA, Reddy KR (2013) Electrokinetic enhanced phytoremediation of soils: status and opportunities. Chemosphere 93:626–636

Carvalho PN, Arias CA, Brix H (2017) Constructed wetlands for water treatment: new developments. Water 9:397

Chai MW, Shi FC, Li RL, Liu FC, Qiu GY, Liu LM (2013) Effect of NaCl on growth and Cd accumulation of halophyte Spartina alterniflora under $CdCl_2$ stress. S Afr J Bot 85:63–69

Chai M, Shi F, Li R, Qiu G, Liu F, Liu L (2014) Growth and physiological responses to copper stress in a halophyte Spartina alterniflora (Poaceae). Acta Physiol Plant 36(3):745–754

Chaiyarat R, Suebsima R, Putwattana N, Kruatrachue M, Pokethitiyook P (2011) Effects of soil amendments on growth and metal uptake by Ocimum gratissimum grown in Cd/Zn-contaminated soil. Water Air Soil Pollut 214(1–4):383–392

Chandra R, Yadav S, Bharagava RN, Murthy RC (2008) Bacterial pretreatment enhances removal of heavy metals during treatment of post-methanated distillery effluent by *Typha angustata* L. J Environ Manag 88(4):1016–1024

Chandra R, Bharagava RN, Yadav S, Mohan D (2009) Accumulation and distribution of toxic metals in wheat (*Triticum aestivum* L.) and Indian mustard (*Brassica campestris* L.) irrigated with distillery and tannery effluents. J Hazard Mater 162(2–3):1514–1521

Chandra R, Saxena G, Kumar V (2015) Phytoremediation of environmental pollutants: an eco-sustainable green technology to environmental management. In: Chandra R (ed) Advances in biodegradation and bioremediation of industrial waste. CRC Press, Boca Raton, pp 1–30. https://doi.org/10.1201/b18218-2

Chaney LR, Scott Angle J, Broadhurst CL, Peters CA, Tappero RV, Sparks DL (2007) Improved understanding of hyperaccumulation yields commercial phytoextraction and phytomining technologies. J Environ Qual 36:1429–1443

Chang F-C, Ko C-H, Tsai M-J, Wang Y-N, Chung C-Y (2014) Phytoremediation of heavy metal contaminated soil by *Jatropha Curcas*. Ecotoxicology 23:1969–1978

Chen B-C, Lai H-Y, Juang K-W (2012) Model evaluation of plant metal content and biomass yield for the phytoextraction of heavy metals by switchgrass. Ecotoxicol Environ Saf 80:393–400

Chen L, Luo S, Li X, Wan Y, Chen J, Liu C (2014) Interaction of Cd hyperaccumulator *Solanum nigrum* L. and functional endophyte *Pseudomonas* sp. Lk9 on soil heavy metals uptake. Soil Biol Biochem 68:300–308

Chirakkara RA, Cameselle C, Reddy KR (2016) Assessing the applicability of phytoremediation of soils with mixed organic and heavy metal contaminants. Rev Environ Sci Biotechnol 15:299–326. https://doi.org/10.1007/s11157-016-9391-0

Cicero-Fernández D, Peña-Fernández M, Expósito-Camargo JA, Antizar-Ladislao B (2015) Role of *Phragmites australis* (common reed) for heavy metals phytoremediation of estuarine sediments. Int J Phytoremediation 18:575–582. https://doi.org/10.1080/15226514.2015.1086306

Conesa MH, Evangelou WHM, Robinson HB et al (2012) A critical view of the current state of phytotechnologies to remediate soils: still a promising tool? Sci World J 2012:173829

Das DK, Chaturvedi OP (2009) Energy dynamics and bioenergy production of *Populus deltoides* G-3 Marsh plantation in eastern India. Biomass Bioenergy 33:144–148

de Souza ER, dos Santos Freire MBG, de Melo DVM, de Assunção Montenegro AA (2014) Management of *Atriplex nummularia* Lindl. in a salt affected soil in a semi arid region of Brazil. Int J Phytoremediation 16:73–85

Dhankher OP, Li Y, Rosen BP, Shi J, Salt D, Senecoff JF, Sashti NA, Meagher RB (2002) Engineering tolerance and hyperaccumulation of arsenic in plants by combining arsenate reductase and γ-glutamylcysteine synthetase expression. Nat Biotechnol 20:1140–1145

Dhankher OP, Pilon-Smits EAH, Meagher RB, Doty S (2011) Biotechnological approaches for phytoremediation. Academic, Oxford, pp 309–328

Dilks RT, Monette F, Glaus M (2016) The major parameters on biomass pyrolysis for hyperaccumulative plants - a review. Chemosphere 146:385–395

Ding Z, Wu J, You A, Huang B, Cao C (2016) Effects of heavy metals on soil microbial community structure and diversity in the rice (*Oryza sativa* L. subsp. Japonica, Food Crops Institute of Jiangsu Academy of Agricultural Sciences) rhizosphere. Soil Sci Plant Nutr 63:75–83

Dinh N, van der Ent A, Mulligan DR, Nguyen AV (2018) Zinc and lead accumulation characteristics and in vivo distribution of Zn2+ in the hyperaccumulator Noccaea caerulescens elucidated with fluorescent probes and laser confocal microscopy. Environ Exp Bot 147:1–12

Dixit P, Singh S, Vanchesswaran R, Patnala K, Eapen S (2010) Expression of a *Neurospora crassa* zinc transporter gene in transgenic *Nicotiana tabacum* enhances plant zinc accumulation without co-transport of cadmium. Plant Cell Environ 35:1696–1707

Dixit R, Wasiullah MD, Pandiyan K, Singh UB, Sahu A, Shukla R, Singh BP, Rai JP, Sharma PK, Lade H, Paul D (2015) Bioremediation of heavy metals from soil and aquatic environment: an overview of principles and criteria of fundamental processes. Sustainability 7:2189–2212

Doni S, Macci C, Peruzzi E, Iannelli R, Masciandaro G (2015) Heavy metal distribution in a sediment phytoremediation system at pilot scale. Ecol Eng 81:146–157

Duman F, Ozturk F (2010) Nickel accumulation and its effect on biomass, protein content and antioxidative enzymes in roots and leaves of watercress (Nasturtium officinale R. Br.). J Environ Sci 22(4):526–532

Ecolotree (2013) Welcome to Ecolotree (Online). http://www.ecolotree.com

Fatima M, Usmani N, Hossain MM, Siddiqui MF, Zafeer MF, Firdaus F, Ahmad S (2014) Assessment of genotoxic induction and deterioration of fish quality in commercial species due to heavy-metal exposure in an urban reservoir. Arch Environ Contam Toxicol 67:203–213

Fellet G, Marchiol L, Perosab D, Zerbiain G (2007) The application of phytoremediation technology in a soil contaminated by pyrite cinders. Ecol Eng 31:207–214

Fruhwirth P, Liebhard P (2004) Miscanthus sinensis Giganteus. Produktion, Inhaltsstoffe und Verwertung. In: Miscanthus sinensis Giganteus: Chi-naachilfals nachwachsender Rohstoff. Landliches Forbildingsinstitut und Landwirtschaftskammer Osterreich, pp 34–38

Gao K, Boiano S, Marzocchella A, Rehmann L (2014) Cellulosic butanol production from alkali-pretreated switchgrass (*Panicum virgatum*) and phragmites (*Phragmites australis*). Bioresour Technol 174:176–181

Garcia-Salgado S, Garcia-Casillas D, Quijano-Nieto MA, Bonilla-Simon MM (2012) Arsenic and heavy metal uptake and accumulation in native plants species from soils polluted by mining activities. Water Air Soil Pollut 223:559–572

Gasic K, Korban SS (2007) Transgenic Indian mustard (Brassica juncea) plants expressing an Arabidopsis phytochelatin synthase (AtPCS1) exhibit enhanced As and Cd tolerance. Plant Mol Biol 64:361

Gerhardt KE, Gerwing PD, Greenberg BM (2016) Opinion: taking phytoremediation from proven technology to accepted practice. Plant Sci 256:170–185. https://doi.org/10.1016/j.plantsci.2016.11.016

Ghosh M, Singh SP (2005) A review on phytoremediation of heavy metals and utilization of its byproducts. Appl Ecol Environ Res 3:1–18

Goel S, Malik JA, Nayyar H (2009) Molecular approach for phytoremediation of metal-contaminated sites. Arch Agron Soil Sci 55:451–475

Gomes HI (2012) Phytoremediation for bioenergy: challenges and opportunities. Environ Technol Rev 1:59–66

Guo J, Dai X, Xu W, Ma M (2008) Overexpressing gsh1 and AsPCS1 simultaneously increases the tolerance and accumulation of cadmium and arsenic in *Arabidopsis thaliana*. Chemosphere 72:1020–1026

Gupta AK, Verma SK, Khan K, Verma RK (2013) Phytoremediation using aromatic plants: a sustainable approach for remediation of heavy metals polluted sites. Environ Sci Technol 47:10115–10116

Habibul N, Hu Y, Wang Y-K, Chen W, Yu H-Q, Sheng G-P (2016) Bioelectrochemical chromium (VI) removal in plant-microbial fuel cells. Environ Sci Technol 50:3882–3889

Hadi F, Bano A (2010) Effect of diazotrophs (*Rhizobium* and *Azobactor*) on growth of maize (*Zea mays* L.) and accumulation of Lead (Pb) in different plant parts. Pak J Bot 42:4363–4370

He CQ, Tan GE, Liang X, Du W, Chen YL, Zhi GY, Zhu Y (2010) Effect of Zn tolerant bacterial strains on growth and Zn accumulation in *Orychophragmus violaceus*. Appl Soil Ecol 44:1–5

He H, Ye Z, Yang D et al (2013) Characterization of endophytic Rahnella sp. JN6 from Polygonum pubescens and its potential in promoting growth and Cd, Pb, Zn uptake by Brassica napus. Chemosphere 90:1960–1965

Ho C-H, Hseu Z-Y, Chen N-C, Tsai C-C (2013) Evaluating heavy metal concentration of plants on a serpentine site for phytoremediation applications. Environ Earth Sci 70:191–199

Howladar SM (2014) A novel Moringa oleifera leaf extract can mitigate the stress effects of salinity and cadmium in bean (*Phaseolus vulgaris* L.) plants. Ecotoxicol Environ Saf 100:69–75

Hsieh J, Chen C, Chiu M, Chein M, Chang J, Endo G et al (2009) Expressing a bacterial mercuric ion binding protein in plant for phytoremediation of heavy metals. J Hazard Mater 161:920–925

Hu Y, Nan Z, Su J, Wang N (2013) Heavy metal accumulation by poplar in calcareous soil with various decrees of multi-metal contamination: implications for phytoextraction and phytostabilization. Environ Sci Pollut Res 20:7194–7203

Huang H, Yu N, Wang L, Gupta DK, He Z, Wang K, Zhu Z, Yan X, Li T, Yang X-e (2011) The phytoremediation potential of bioenergy crop *Ricinus communis* for DDTs and cadmium co-contaminated soil. Bioresour Technol 102:11034–11038

Huang JC, Suárez MC, Yang SI, Lin Z-Q, Terry N (2013) Development of a constructed wetland water treatment system for selenium removal: incorporation of an algal treatment component. Environ Sci Technol 47(18):10518–10525

Huang J, Liu Z, Li S, Xu B, Gong Y, Yang Y, Sun H (2016) Isolation and engineering of plant growth promoting rhizobacteria *Pseudomonas aeruginosa* for enhanced cadmium bioremediation. J Gen Appl Microbiol 62:258–265

Ike A, Sriprang R, Ono H, Murooka Y, Yamashita M (2007) Bioremediation of cadmium contaminated soil using symbiosis between leguminous plant and recombinant rhizobia with the *MTL4* and the *PCS* genes. Chemosphere 66:1670–1676

Ike A, Sriprang R, Ono H, Murooka Y, Yamashita M (2008) Promotion of metal accumulation in nodule of *Astragalus sinicus* by the expression of the iron-regulated transporter gene in *Mesorhizobium huakuii* subsp. *rengei* B3. J Biosci Bioeng 105:642–648

Jacob JM, Karthik C, Saratale RG, Kumar SS, Prabakar D, Kadirvelu K, Pugazhendhi A (2018) Biological approaches to tackle heavy metal pollution: a survey of literature. J Environ Manag 217:56–70

Jaishankar M, Tseten T, Abbalagan N, Mathew BB, Beeregowda KN (2014) Toxicity, mechanism and health effects of some heavy metals. Interdiscip Toxicol 7:60–72

Jan FA, Ishaq M, Khan S et al (2010) A comparative study of human health risks via consumption of food crops grown on wastewater irrigated soil (Peshawar) and relatively clean water irrigated soil (lower Dir). J Hazard Mater 179(1–3):612–621

Javed M, Usmani N (2015) Impact of heavy metal toxicity on hematology and glycogen status of fish: a review. Proc Natl Acad Sci India Sect B Biol Sci 85:889–900

Javed M, Ahmad I, Usmani N, Ahmad M (2016) Bioaccumulation, oxidative stress and genotoxicity in fish (*Channa punctatus*) exposed to a thermal power plant effluent. Ecotoxicol Environ Saf 127:163–169

Jezierska B, Ługowska K, Witeska M (2009) The effects of heavy metals on embryonic development of fish (a review). Fish Physiol Biochem 35(4):625–640

Jiang Y, Lei M, Duan L, Longhurst P (2015) Integrating phytoremediation with biomass valorisation and critical element recovery: a UK contaminated land perspective. Biomass Bioenergy 83:328–339

Jianxu W, Xinbin F, Christopher WNA, Ying X, Lihai S (2012) Remediation of mercury contaminated sites - a review. J Hazard Mater 221–222:1–18

Jing YX, Yan JL, He HD, Yang DJ, Xiao L, Zhong T, Yuan M, Cai XD, Li SB (2014) Characterization of bacteria in the rhizosphere soils of *polygonum pubescens* and their potential in promoting growth and Cd, Pb, Zn uptake by *Brassica napus*. Int J Phytoremediation 16:321–333

Kale RA, Lokhande VH, Ade AB (2015) Investigation of chromium phytoremediation and tolerance capacity of a weed. Portulaca oleracea L. in a hydroponic system. Water Environ J 29(2):236–242

Kalve S, Sarangi BK, Pandey RA, Chakrabarti T (2011) Arsenic and chromium hyperaccumulation by an ecotype of *Pteris vittata*-prospective for phytoextraction from contaminated water and soil. Curr Sci 100:888–894

Khan S, Cao Q, Zheng YM, Huang YZ, Zhu YG (2008) Health risks of heavy metals in contaminated soils and food crops irrigated with wastewater in Beijing, China. Environ Pollut 152:686–692

Khan S, Ahmad I, Shah MT, Rehman S, Khaliq A (2009) Use of constructed wetland for the removal of heavy metals from industrial wastewater. J Environ Manage 90:3451–3457

Khan A, Sharif M, Ali A, Shah SNM, Mian IA, Wahid F, Jan B, Adnan M, Nawaz S, Ali N (2014) Potential of AM fungi in phytoremediation of heavy metals and effect on yield of wheat crop. Am J Plant Sci 5:1578–1586

Khaokaew S, Landrot G (2014) A field-scale study of cadmium phytoremediation in a contaminated agricultural soil at Mae Sot District, Tak Province, Thailand: (1) Determination of Cd-hyperaccumulating plants. Chemosphere 138:883–887. https://doi.org/10.1016/j.chemosphere.2014.09.108

Kim SJ, Kim MY, Jeong SJ, Jang MS, Chung IM (2012) Analysis of the biomass content of various *Miscanthus* genotypes for biofuel production in Korea. Ind Crops Prod 38:46–49

Koptsik GN (2014) Problems and prospects concerning the phytoremediation of heavy metal polluted soils: a review. Eurasian Soil Sci 47:923–939

Kotrba P, Najmanova J, Macek T, Ruml T, Mackova M (2009) Genetically modified plants in phytoremediation of heavy metal and metalloid soil and sediment pollution. Biotechnol Adv 27(6):799–810

Lajayar BA, Ghorbanpourb M, Nikabadi S (2017) Heavy metals in contaminated environment: destiny of secondary metabolite biosynthesis, oxidative status and phytoextraction in medicinal plants. Ecotoxicol Environ Saf 145:377–390

Lavanya C, Murthy IYLL, Nagaraj G, Mukta N (2012) Prospects of castor (*Ricinus communis* L.) genotypes for biodiesel production in India. Biomass Bioenergy 39:204–209

Lee JH (2013) An overview of phytoremediation as a potentially promising technology for environmental pollution control. Biotechnol Bioprocess Eng 18:431–439

Lefèvre I, Marchal G, Edmond GM, Correal E, Lutts S (2010) Cadmium has contrasting effects on polyethylene glycol - sensitive and resistant cell lines in the Mediterranean halophyte species *Atriplex halimus* L. J Plant Physiol 167(5):365–374

Li C, Xu Y, Jiang W, Dong X, Wang D, Liu B (2013) Effect of NaCl on the heavy metal tolerance and bioaccumulation of Zygosaccharomyces rouxii and Saccharomyces cerevisiae. Bioresour Technol 143:46–52

Li X, Li D, Yana Z, Ao Y (2018) Biosorption and bioaccumulation characteristics of cadmium by plant growth-promoting rhizobacteria. RSC Adv 8:30902–30911

Liang L, Liu W, Sun Y, Huo X, Li S, Zhou Q (2017) Phytoremediation of heavy metal contaminated saline soils using halophytes: current progress and future perspectives. Environ Rev 25:269–281

Lindsay (1979) Permissible limits for metals. http://www.occeweb.com/og/metals-limits.pdf

Liu D, An Z, Mao Z, Ma L, Lu Z (2015) Enhanced heavy metal tolerance and accumulation by transgenic sugar beets expressing Streptococcus thermophilus StGCS-GS in the presence of Cd, Zn and Cu alone or in combination. PLoS One 10(6):e0128824

Luis R, Rincon J, Asencio I, Rodriguez-Castellanos L (2007) Capability of selected crop plants for shoot mercury accumulation from polluted soils: phytoremediation perspectives. Int J Phytoremediation 9:1–13

Luo S, Chen L, Chen J, Xiao X, Xu T, Wan Y, Rao C, Liu C, Liu Y, Lai C, Zeng G (2011) Analysis and characterization of cultivable heavy metal-resistant bacterial endophytes isolated from Cd hyperaccumulator *Solanum nigrum* L. and their potential use for phytoremediation. Chemosphere 85:1130–1138

Łuszczek-Trojnar E, Drąg-Kozak E, Szczerbik P, Socha M, Popek W (2014) Effect of long-term dietary lead exposure on some maturation and reproductive parameters of a female Prussian carp (*Carassius gibelio* B.). Environ Sci Pollut Res Int 21:2465–2478

Ma Y, Prasad MNV, Rajkumar M, Freitas H (2011) Plant growth promoting rhizobacteria and endophytes accelerate phytoremediation of metalliferous soils. Biotechnol Adv 29:248–258

Ma Y, Oliviera RS, Nai F, Rajkumar M, Luo Y, Rocha I, Freitas H (2015) The hyperaccumulator *Sedum plumbizincicola* harbors metal-resistant endophytic bacteria that improve its phytoextraction capacity in multi-metal contaminated soil. J Environ Manag 156:62–69

Ma Y, Rajkumar M, Zhang C, Freitas H (2016a) Inoculation of *Brassica oxyrrhina* with plant growth promoting bacteria for the improvement of heavy metal phytoremediation under drought conditions. J Hazard Mater 320:36–44

Ma Y, Oliveira RS, Freitas H, Zhang C (2016b) Biochemical and molecular mechanisms of plant-microbe-metal interactions: relevance for phytoremediation. Front Plant Sci 7:918

Maestri E, Marmiroli N (2011) Transgenic plants for phytoremediation. Int J Phytoremediation 13:264–279

Mahar A, Wang P, Ali A, Awasthi MK, Lahori AH, Wang Q, Li R, Zhang Z (2016) Challenges and opportunities in the phytoremediation of heavy metals contaminated soils: a review. Ecotoxicol Environ Saf 26:111–121

Mahmood Q, Pervez A, Zeb BS et al (2013) Natural treatment systems as sustainable eco-technologies for the developing countries. Biomed Res Int 2013:796373

Manousaki E, Kalogerakis N (2011) Halophytes - an emerging trend in phytoremediation. Int J Phytoremediation 13:959–969

Manousaki E, Kadukova J, Papadantonakis N, Kalogerakis N (2008) Phytoextraction and phytoexcretion of Cd by *Tamarix smyrnensis* growing on contaminated non saline and saline soils. Environ Res 106:326–332

Mao X, Han FX, Shao X, Guo K, McComb J, Arslan Z, Zhang Z (2016) Electro-kinetic remediation coupled with phytoremediation to remove lead, arsenic and cesium from contaminated paddy soil. Ecotoxicol Environ Saf 125:16–24

Marchiol L, Fellet G, Perosa D, Zerbi G (2007) Removal of trace metals by *Sorghum bicolor* and *Helianthus annuus* in a site polluted by industrial wastes: a field experience. Plant Physiol Biochem 45:379–387

Marques APGC, Rangel AOSS, Castro PML (2009) Remediation of heavy metal contaminated soils: phytoremediation as a potentially promising clean-up technology. Crit Rev Environ Sci Technol 39:622–654

Maszenan AM, Liu Y, Ng WJ (2011) Bioremediation of wastewaters with recalcitrant organic compounds and metals by aerobic granules. Biotechnol Adv 29(1):111–123

Meers E, Van Slycken S, Adriaensen K, Ruttens A, Vangronsveld J, Laing GD, Witters N, Thewys T, Tack FMG (2010) The use of bio-energy crops (*Zea mays*) for phytoattenuation of heavy metals on moderately contaminated soils: a field experiment. Chemosphere 78:35–41

Meier S, Borie F, Bolan N, Cornejo P (2012) Phytoremediation of metal-polluted soils by arbuscular mycorrhizal fungi. Crit Rev Environ Sci Technol 42(7):741–775

Mekawy AMM, Assaha DVM, Munehiro R et al (2018) Characterization of type 3 metallothionein-like gene (OsMT-3a) from rice, revealed its ability to confer tolerance to salinity and heavy metal stresses. Environ Exp Bot 147:157–166

Mishra S, Bharagava RN (2016) Toxic and genotoxic effects of hexavalent chromium in environment and its bioremediation strategies. J Environ Sci Health C Environ Carcinog Ecotoxicol Rev 34:1–32

Mleczek M, Rutkowski P, Rissmann I, Kaczmarek Z, Golinski P, Szentner K, Strazynska K, Stachowiak A (2010) Biomass productivity and phytoremediation potential of *Salix alba* and *Salix viminalis*. Biomass Bioenergy 34:1410–1418

Mukherjee G, Saha C, Naskar N et al (2018) An Endophytic Bacterial Consortium modulates multiple strategies to improve Arsenic Phytoremediation Efficacy in Solanum nigrum. Sci Rep 8:6979

Mukhopadhyay S, Maiti SK (2010) Phytoremediation of metal mine waste. Appl Ecol Environ Res 8:207–222

Muthusaravanan S, Sivarajasekar N, Vivek JS, Paramasivan T, Naushad M, Prakashmaran J, Gayathri V, Al-Duaij OK (2018) Phytoremediation of heavy metals: mechanisms, methods and enhancements. Environ Chem Lett 16:1339–1359

Nagajyoti PC, Lee KD, Sreekanth TVM (2010) Heavy metals, occurrence and toxicity for plants. Environ Chem Lett 8:199–216

Nahar N, Aminur R, Nawani NN, Ghosh S, Mandal A (2017) Phytoremediation of arsenic from the contaminated soil using transgenic tobacco plants expressing ACR2 gene of *Arabidopsis thaliana*. J Plant Physiol 218:121–126. https://doi.org/10.1016/j.jplph.2017.08.001

Nazeer S, Hashmi MZ, Malik RN (2014) Heavy metals distribution, risk assessment and water quality characterization by water quality index of the River Soan, Pakistan. Ecol Indic 43:262–270

Nie L, Shah S, Rashid A (2002) Phytoremediation of arsenate contaminated soil by transgenic canola and the plant growth-promoting bacterium Enterobacter cloacae CAL2. Plant Physiol Biochem 40:355–361

Nirola R, Megharaj M, Palanisami T, Aryal R, Venkateswarlu K, Naidu R (2015) Evaluation of metal uptake factors of native trees colonizing an abandoned copper mine: a quest for phytostabilization. J Sustain Mining 14:115–123

Nunes da Silva M, Mucha AP, Rocha AC, Silva C, Carli C, Gomes CR, Almeida CMR (2014) Evaluation of the ability of two plants for the phytoremediation of Cd in salt marshes. Estuar Coast Shelf Sci 141:78–84

Oosten MJ, Van Maggio A (2015) Functional biology of halophytes in the phytoremediation of heavy metal contaminated soils. Environ Exp Bot 11:135–146

Pandey VC, Bajpai O, Singh N (2016) Energy crops in sustainable phytoremediation. Renew Sust Energ Rev 54:58–73

Park JH, Lamba D, Paneerselvama P, Choppalaa G, Bolana N, Chung J-W (2011) Role of organic amendments on enhanced bioremediation of heavy metal(loid) contaminated soils. J Hazard Mater 185:549–574

Paz-Ferreiro J, Lu H, Fu S, Méndez A, Gascó G (2014) Use of phytoremediation and biochar to remediate heavy metal polluted soils: a review. Solid Earth 5:65–75

Pedro CA, Santos MSS, Ferreira SMF, Gonçalves SC (2013) The influence of cadmium 1051 contamination and salinity on the survival, growth and phytoremediation capacity of the saltmarsh plant Salicornia ramosissima. Mar Environ Res 92:197–205

Perez-Cruzado C, Merino A, Rodnguez-Soalleiro R (2011) A management tool for estimating bioenergy production and carbon sequestration in *Eucalyptus globulus* and *Eucalyptus nitens* grown as short rotation woody crops in north-west Spain. Biomass Bioenergy 35:2839–2851

Ping Z, Wensheng S, Zhian L, Bin L, Jintian L, Jingsong S (2009) Removal of metals by sorghum plants from contaminated land. J Environ Sci 21:1432–1437

Prade T, Svensson S-E, Mattsson JE (2012) Energy balances for biogas and solid biofuel production from industrial hemp. Biomass Bioenergy 40:36–52

Prasad A, Singh AK, Chand S, Chanotiya C, Patra D (2010) Effect of chromium and lead on yield, chemical composition of essential oil, and accumulation of heavy metals of mint species. Commun Soil Sci Plant Anal 41(18):2170–2186

Quievryn G, Peterson E, Messer J, Zhitkovich A (2003) Genotoxicity and mutagenicity of chromium(VI)/ascorbate-generated DNA adducts in human and bacterial cells. Biochemist 42(4):1062–1070

Rai PK (2008) Phytoremediation of Hg and Cd from industrial effluents using an aquatic free-floating macrophyte *Azolla pinnata*. Int J Phytoremediation 10:430–439

Rai R, Pandey S, Rai SP (2011) Arsenic-induced changes in morphological, physiological, and biochemical attributes and artemisinin biosynthesis in Artemisia annua, an antimalarial plant. Ecotoxicology 20:1900–1913

Rajkumar M, Sandhya S, Prasad MN, Freitas H (2012) Perspectives of plant-associated microbes in heavy metal phytoremediation. Biotechnol Adv 30:1562–1574

Ramamurthy AS, Memarian R (2014) Chelate enhanced phytoremediation of soil containing a mixed contaminant. Environ Earth Sci 72:201

Reddy KR, Amaya-Santos G, Yargicoglu E, Cooper DE, Negri MC (2017) Phytoremediation of heavy metals and PAHs at slag fill site: three-year field-scale investigation. Int J Geotech Eng 13(1):32–47. https://doi.org/10.1080/19386362.2017.1318231

Redfern LK, Gunsch CK (2016) Endophytic phytoaugmentation: treating wastewater and runoff through augmented phytoremediation. Ind Biotechnol (New Rochelle N Y) 12:83–90

Redondo-Gómez S, Mateos-Naranjo E, Andrades-Moreno L (2010) Accumulation and tolerance 1074 characteristics of cadmium in a halophytic Cd-hyperaccumulator, Arthrocnemum macrostachyum. J Hazard Mater 184(1–3):299–307

Reeves RD, Baker AJM (2000) Metal-accumulating plants. In: Raskin I, Ensley BD (eds) Phytoremediation of toxic metals: using plants to clean-up the environment. Wiley, New York, pp 193–230

Robinson B, Green S, Mills T et al (2003) Phytoremediation: using plants as biopumps to improve degraded environments. Aust J Soil Res 41:599–611

Sahu RK, Katiyar S, Tiwari J, Kisku GC (2007) Assessment of drain water receiving effluent from tanneries and its impact on soil and plants with particular emphasis on bioaccumulation of heavy metals. J Environ Biol 28(3):685–690

Sakakibara M, Ohmori Y, Ha NTH, Sano S, Sera K (2011) Phytoremediation of heavy metal contaminated water and sediment by *Eleocharis acicularis*. Clean Soil Air Water 39:735–741

Salazar MJ, Pignata ML (2014) Lead accumulation in plants grown in polluted soils. Screening of native species for phytoremediation. J Geochem Explor 137:29–36

Salihaj M, Bani A, Echevarria G (2016) Heavy metals uptake by hyperaccumulating flora in some serpentine soils of Kosovo. Global NEST J 18:214–222

Sarwar N, Imran M, Shaheen MR, Ishaq W, Kamran A, Matloob A, Rehim A, Hussain S (2017) Phytoremediation strategies for soils contaminated with heavy metals: modifications and future perspectives. Chemosphere 171:710–721

Saxena G, Bharagava RN (2017) Organic and inorganic pollutants in industrial wastes, their ecotoxicological effects, health hazards and bioremediation approaches. In: Bharagava RN (ed) Environmental pollutants and their bioremediation approaches, 1st edn. CRC Press, Boca Raton, pp 23–56. https://doi.org/10.1201/9781315173351-3

Saxena G, Chandra R, Bharagava RN (2016) Environmental pollution, toxicity profile and treatment approaches for tannery wastewater and its chemical pollutants. Rev Environ Contam Toxicol 240:31–69

Segura R, Arancibia V, Zuniga MC, Pasten P (2006) Distribution of copper, zinc, lead and cadmium concentrations in stream sediments from the Mapocho River in Santiago, Chile. J Geochem Explor 91:71–80

Sheoran V, Sheoran AS, Poonia P (2009) Phytomining: a review. Min Eng 22:1007–1019

Shim D, Kim S Choi Y-I, Song W-Y, Park P, Youk ES, Jeong S-C, Martinoia E, Noh E-W, Lee Y (2013) Transgenic poplar trees expressing yeast cadmium factor 1 exhibit the characteristics necessary for the phytoremediation of mine tailing soil. Chemosphere 90:1478–1486

Siddiqui F, Krishna SK, Tandon P, Srivastava S (2013) Arsenic accumulation in Ocimum spp. and its effect on growth and oil constituents. Acta Physiol Plant 35(4):1071–1079

Simpson JA, Picchi G, Gordon AM, Thevathasan NV, Stanturf J, Nicholas I (2009) Short rotation crops for bioenergy systems. Environmental benefits associated with short-rotation woody crops, Technical Review No. 3, IEA Bioenergy. http://www.shortrotationcrops.org/taskreports.htm

Singh S, Korripally P, Vancheeswaran R, Eapen S (2011) Transgenic *Nicotiana tabacum* plants expressing a fungal copper transporter gene show enhanced acquisition of copper. Plant Cell Rep 30:1929–1938

Sobolev D, Begonia MFT (2008) Effects of heavy metal contamination upon soil microbes: lead-induced changes in general and denitrifying microbial communities as evidenced by molecular markers. Int J Environ Res Public Health 5:450–456

Sone Y, Nakamura R, Pan-Hou H, Sato MH, Itoh I, Kiyono M (2013) Increase methylmercury accumulation in *Arabidopsis thaliana* expressing bacterial broad-spectrum mercury transporter MerE. AMB Express 3:52

Sriprang R, Hayashi M, Yamashita M, Ono H, Saeki K, Murooka Y (2002) A novel bioremediation system for heavy metals using the symbiosis between leguminous plant and genetically engineered rhizobia. J Biotechnol 99:279–293

Sriprang R, Hayashi M, Ono H, Takagi M, Hirata K, Murooka Y (2003) Enhanced accumulation of Cd2+ by a Mesorhizobium sp. transformed with a gene from Arabidopsis thaliana coding for phytochelatin synthase. Appl Environ Microbiol 69:1791–1796

Srivastava S, Verma PC, Chaudhary V, Singh N, Abhilash PC, Kumar KV, Sharma N, Singh N (2013) Inoculation of arsenic-resistant *Staphylococcus arlettae* on growth and arsenic uptake in *Brassica juncea* (L.) Czern. Var. R-46. J Hazard Mater 262:1039–1047

Stephenson C, Black CR (2014) One step forward, two steps back: the evolution of phytoremediation into commercial technologies. Biosci Horizons 7:1–15

Stottmeister U, Wießner A, Kuschk P, Kappelmeyer U, Kastner M, Bederski O, Muller RA, Moormann H (2003) Effects of plants and microorganisms in constructed wetlands for wastewater treatment. Biotechnol Adv 22:93–117

Sun YY, Xu HX, Li JH, Shi XQ, Wu JC, Ji R, Guo HY (2015) Phytoremediation of soils contaminated with phenanthrene and cadmium by growing Willow (Salix × aureopendula CL ‘J1011’). Int J Phytoremediation 18(2):150–156

Tiwari S, Singh SN, Garg SK (2012) Stimulated phytoextraction of metals from fly ash by microbial interventions. Environ Technol 33:2405–2413

Tomović GM, Mihailović ML, Tumi AF, Gajić BA, Mišljenović TD, Niketić MS (2013) Trace metals in soils and several Brassicaceae plant species from serpentine sites of Serbia. Arch Environ Prot 39:29–49

Ullah A, Heng S, Munis MFH, Fahad S, Yang X (2015) Phytoremediation of heavy metals assisted by plant growth promoting (PGP) bacteria: a review. Environ Exp Bot 117:28–40

Vamerali T, Bandiera M, Coletto L, Zanetti F, Dickinson NM, Mosca G (2009) Phytoremediation trials on metal- and arsenic-contaminated pyrite wastes (Torviscosa, Italy). Environ Pollut 157:887–894

Vamerali T, Marchiol L, Bandiera M, Fellet G, Dickinson N, Lucchini P, Mosca G, Zerbi G (2012) Advances in agronomic management of phytoremediation: Methods and results from a 10-year study of metal-polluted soils. Ital J Agron 7(4):e42. https://doi.org/10.4081/ija.2012.e42

Vangronsveld J, Herzig R, Weyens N, Boulet J, Adriaensen K, Ruttens A, Thewys T, Vassilev A, Meers E, Nehnevajova E, Van der Lelie D, Mench M (2009) Phytoremediation of contaminated soils and groundwater: lessons from the field. Environ Sci Pollut Res 16:765–794

Venkateswaran P, Vellaichamy S, Palanivelu K (2007) Speciation of heavy metals in electroplating industry sludge and wastewater residue using inductively coupled plasma. Int J Environ Sci Technol 4(4):497–504

Wan X, Lei M, Chen T (2016) Cost–benefit calculation of phytoremediation technology for heavy-metal-contaminated soil. Sci Total Environ 563–564:796–802

Wang H, Zhong G (2011) Effect of organic ligands on accumulation of copper in hyperaccumulator and nonaccumulator Commelina communis. Biol Trace Elem Res 143(1):489–499

Wang Y-C, Ko C-H, Chang F-C, Chen P-Y, Liu T-F, Sheu Y-S, Shih T-L, Teng C-J (2011) Bioenergy production potential for aboveground biomass from a subtropical constructed wetland. Biomass Bioenergy 35:50–58

Wang H-L, Tian C-Y, Jiang L, Wang L (2014) Remediation of heavy metals contaminated\ saline soils: a halophyte choice? Environ Sci Technol 48:21–22
Weggler-Beaton K, McLaughlin MJ, Graham RD (2000) Salinity increases cadmium uptake by wheat and Swiss chard from soil amended with biosolids. Aus J Soil Res 38(1):37–45
Wei S, da Silva JAT, Zhou Q (2008) Agro-improving method of phytoextracting heavy metal contaminated soil. J Hazard Mater 150:662–668
Weyens N, Schellingen K, Beckers B, Janssen J, Ceulemans R, van der Lelie D, Taghavi CR, Vangronsveld J (2013) Potential of willow and its genetically engineered associated bacteria to remediate mixed Cd and toluene contamination. J Soils Sediments 13:176–188
Willscher S, Mirgorodsky D, Jablonski L, Ollivier D, Merten D, Büchel G, Wittig J, Werner P (2013) Field scale phytoremediation experiments on a heavy metal and uranium contaminated site, and further utilization of the plant residues. Hydrometallurgy 131–132:46–53
Wiszniewska A, Hanus-Fajerska E, Muszynska E, Ciarkowska K (2016) Natural organic amendments for improved phytoremediation of polluted soils: a review of recent progress. Pedosphere 26:1–12
Wu CH, Wood TK, Mulchandani A, Chen W (2006) Engineering plant-microbe symbiosis for rhizoremediation of heavy metals. Appl Environ Microbiol 72:1129–1134
Wu G, Kanga H, Zhang X, Shaob H, Chuc K, Ruan CA (2010) A critical review on the bio-removal of hazardous heavy metals from contaminated soils: issues, progress, eco-environmental concerns and opportunities. J Hazard Mater 174:1–8
Wu H, Liu X, Zhao J, Yu J (2013) Regulation of metabolites, gene expression, and antioxidant enzymes to environmentally relevant lead and zinc in the halophyte Suaeda salsa. J Plant Growth Regul 32(2):353–361
Wuana RA, Okieimen FE (2011) Heavy metals in contaminated soils: a review of sources, chemistry, risks and best available strategies for remediation. ISRN Ecol 2011:402647. https://doi.org/10.5402/2011/402647
Xia Y, Liu J, Wang Y, Zhang XX, Shen ZG, Hu Z et al (2018) Ectopic expression of Vicia sativa Caffeoyl-CoA O-methyltransferase (VsCCoAOMT) increases the uptake and tolerance of cadmium in Arabidopsis. Environ Exp Bot 145:47–53
Xie Z, Wu L, Chen N, Liu C, Zheng Y, Xu S, Li F, Xu Y (2012) Phytoextraction of Pb and Cu contaminated soil with maize and Microencapsulated EDTA. Int J Phytoremediation 14:727–740
Xu J, Yin HX, Liu X, Li X (2010) Salt affects plant Cd-stress responses by modulating growth and Cd accumulation. Planta 231(2):449–459
Yadav SK (2010) Heavy metals toxicity in plants: an overview on the role of glutathione and phytochelatins in heavy metal stress tolerance of plants. S Afr J Bot 76:167–179
Yadava KK, Gupta N, Kumar A, Reece LM, Singh N, Rezania S, Khan SA (2018) Mechanistic understanding and holistic approach of phytoremediation: a review on application and future prospects. Ecol Eng 120:274–298
Yong X, Chen Y, Liu W, Xu L, Zhou J, Wang S, Chen P, Ouyang P, Zheng T (2014) Enhanced cadmium resistance and accumulation in Pseudomonas putida KT2440 expressing the phytochelatin synthase gene of Schizosaccharomyces pombe. Lett Appl Microbiol 58:255–261
Yuan M, He H, Xiao L, Zhong T, Liu H, Li S, Deng P, Ye Z, Jing Y (2013) Enhancement of Cd phytoextraction by two Amaranthus species with endophytic *Rahnella* sp. JN27. Chemosphere 103:99–104
Zhan Y, Liu J, Zhou Y et al (2013) enhanced phytoremediation of mixed heavy metal (mercury)–organic pollutants (trichloroethylene) with transgenic alfalfa co-expressing glutathione S-transferase and human P450 2E1. J Hazard Mater 15:1100–1107
Zhang X, Xia H, Li Z, Zhuang P, Gao B (2011) Identification of a new potential Cd-hyperaccumulator *Solanum photeinocarpum* by soil seed bank-metal concentration gradient method. J Hazard Mater 189:414–419
Zhang D-Q, Jinadasa KBSN, Gersberg RM, Liu Y, Ng WJ, Tan SK (2014) Application of constructed wetlands for wastewater treatment in developing countries - a review of recent developments (2000–2013). J Environ Manag 141:116–131

Zhang D-Q, Jinadasa KBSN, Gersberg RM, Liu Y, Tan SK, Ng WJ (2015a) Application of constructed wetlands for wastewater treatment in tropical and subtropical regions (2000-2013). J Environ Sci 30:30–46
Zhang H, Chen X, He C, Liang X, Oh K, Liu X, Lei Y (2015b) Use of energy crop (*Ricinus communis* L.) for phytoextraction of heavy metals assisted with citric acid. Int J Phytoremediation 17:632–639
Zhuang X, Chen J, Shim H, Zhihui B (2007) New advances in plant growth-promoting rhizobacteria for bioremediation. Environ Int 33:406–413
Zhang J, Zhang M, Shohag MJ et al (2016) Enhanced expression of SaHMA3 plays critical roles in Cd hyperaccumulation and hypertolerance in Cd hyperaccumulator Sedum alfredii Hance. Planta 243(3):577–589
Zucaro A, Forte A, Basosi R, Fagnano M, Fierro A (2016) Life cycle assessment of second generation bioethanol produced from low-input dedicated crops of Arundo donax L. Bioresour Technol 219:589–599

Snakes as Novel Biomarkers of Mercury Contamination: A Review

David L. Haskins, Robert M. Gogal Jr., and Tracey D. Tuberville

Contents

D. L. Haskins (✉)
Savannah River Ecology Laboratory, University of Georgia, Aiken, SC, USA

Interdisciplinary Toxicology Program, University of Georgia, Athens, GA, USA

Warnell School of Forestry and Natural Resources, University of Georgia, Athens, GA, USA
e-mail: david.haskins@uga.edu

R. M. Gogal Jr.
Department of Biosciences and Diagnostic Imaging, College of Veterinary Medicine, University of Georgia, Athens, GA, USA
e-mail: rgogal@uga.edu

T. D. Tuberville
Savannah River Ecology Laboratory, University of Georgia, Aiken, SC, USA
e-mail: tubervil@uga.edu

P. de Voogt (ed.), *Reviews of Environmental Contamination and Toxicology, Volume 249*, Reviews of Environmental Contamination and Toxicology Volume 249, https://doi.org/10.1007/398_2019_26

1 Introduction

1.1 Snakes, Ecotoxicology, and Mercury

Ecological risk assessments are a tool used by regulatory agencies to determine how the environment and associated wildlife might be impacted by proposed or ongoing anthropogenic activities (Newman 2015). In the last two decades, researchers have highlighted the absence of reptiles in these assessments (Campbell and Campbell 2001; Grillitsch and Schiesari 2010; Weir et al. 2010). Reptiles are often underrepresented or excluded from ecological risk assessments even though they serve important roles in their ecosystems and can be both predators and prey of terrestrial and aquatic species. Furthermore, reptiles often inhabit environments that are conducive to long-term retention of contaminants (e.g., methylation of mercury in isolated wetlands), increasing the likelihood that they will be exposed to and accumulate pollutants. Physiological characteristics found in reptiles also make them ideal study candidates for risk assessments. Their metabolic mode (ectothermy) allows high tissue conversion rates and low tissue turnover rates, which may promote processes associated with bioaccumulation (Hopkins 2006). Many of these organisms also exhibit indeterminate growth, and this may allow researchers to quantify bioaccumulation rates over their entire lifespan.

Although interest in reptile ecotoxicology has increased in recent years, snakes are still widely excluded from such studies, even though their life history characteristics make them useful candidate receptor species (Hopkins 2000; Campbell and Campbell 2001; Drewett et al. 2013). This is likely due to a combination of factors. For instance, snake studies have historically been plagued by small sample sizes, and thus, accumulation of information regarding their life history and ecological traits has lagged behind other reptiles. Many species are also secretive, limiting their detection or capture (Durso et al. 2011; Willson and Winne 2016; Willson et al. 2011). Because of these issues, research involving these species may be more difficult to conduct and interpret. Cultural biases toward snakes are also often extreme, and people tend to either worship or loathe these reptiles (Pough et al. 1998; Campbell and Campbell 2001). Further, it is often difficult to convince funding agencies and the general public that results from snake studies positively impact humans, although environmental education efforts may be helping in this regard (Shine and Bonnet 2000). Overall, it is no surprise that research in reptile ecotoxicology has significantly lagged behind that of other taxonomic groups (Hopkins 2000; Weir et al. 2010).

There are more than 3,000 species of snakes recognized globally, and they exhibit a wide diversity in their habitat associations. These limbless predators inhabit every continent except Antarctica and have successfully colonized terrestrial, freshwater, and marine habitats (Vitt and Caldwell 2014). Of the snake families, Colubridae is the most speciose and is comprised of more than half of the world's described snake species (>1,700 species, Vitt and Caldwell 2014). Although the ICUN Red List of Threatened Species does not indicate that large numbers of snake species are threatened (~185 at risk, IUCN 2017), it is notoriously difficult to study snake population dynamics, and recent research has suggested that snakes (like other reptiles) could be experiencing population declines likely due to an assortment of threats, including pollution (Gibbons et al. 2000; Reading et al. 2010).

Mercury (Hg) is a major contaminant of concern around the globe, and its prevalence has increased markedly in both terrestrial and aquatic habitats over the last half a century (Tweedy et al. 2013; Lamborg et al. 2014). Anthropogenic activities such as mining, coal combustion, and other industrial processes have facilitated Hg's release into the environment, where it is transformed into a bioavailable form (methylmercury, MeHg) that can accumulate in wildlife (Schneider et al. 2013). Recent studies suggest that snakes can accumulate significant amounts of Hg, with snakes inhabiting aquatic environments often exhibiting the highest Hg burdens (Axelrad et al. 2011; Drewett et al. 2013). Indeed, most ecotoxicological studies of Hg bioaccumulation in snakes have focused on aquatic or semiaquatic species such as cottonmouths (*Agkistrodon piscivorus*), natricines (e.g., *Nerodia* and *Thamnophis* spp.), and a few other colubrids (Campbell and Campbell 2001; Schneider et al. 2013; Lemaire et al. 2018).

In this review, we will examine the sources of Hg contamination and the factors that determine availability within aquatic and terrestrial environments. We will also review the potential of snakes as biomarkers for Hg exposure, focusing on bioaccumulation and known effects of exposure. We will then discuss the role snakes play in nutrient and contaminant transfer. Finally, we will conclude by identifying research gaps.

2 Mercury

2.1 Mercury Sources

Mercury is a highly toxic heavy metal that is released into the environment by natural and anthropogenic sources. Unlike other heavy metals of concern (e.g., lead or cadmium), Hg is unique not only because is it locally transformed to more bioavailable forms (e.g., Hg methylation), but because it also can be transported on a global scale via atmospheric cycling and deposition (Boening 2000; Zhang et al. 2009). In fact, distinct atmospheric cycles may result in a system where regions can be impacted by Hg that originates from both local and global inputs. Prior to the industrial revolution, Hg in the environment mostly originated from the natural mobilization of Hg deposits in the earth's crust and volcanoes (Selin 2009). Presently, environmental Hg originates from a variety of anthropogenic activities, including coal burning, mining, water treatment plants, and other industrial facilities (Selin 2009). Thus, industrialization has significantly altered Hg emissions and cycling. For example, one study recently suggested that Hg concentrations in ocean surface waters have tripled compared to pre-anthropogenic conditions (Lamborg et al. 2014).

2.2 *Forms of Mercury and Availability in the Environment*

The major forms of Hg in the environment include elemental, ionic, or organic Hg (e.g., methylmercury or MeHg). Atmospheric Hg is primarily comprised of elemental Hg, which has a long atmospheric lifespan (~1 year) and thus contributes to its global transport (Chen et al. 2014). Within aquatic and terrestrial habitats, Hg can be found in various forms, but most studies focus on inorganic Hg and MeHg. The most toxic form of Hg, MeHg, is commonly produced by interactions between anaerobic sulfur-reducing bacteria and inorganic Hg (Klaus et al. 2016). Bioaccumulation of MeHg is facilitated by dietary exposure, with higher Hg burdens reported in top trophic predators (Scheuhammer et al. 2007).

MeHg deposition in aquatic environments is dependent on the original Hg species present, as well as an assortment of environmental factors (e.g., pH, temperature, oxygen, dissolved organic carbon, sediment type, forest cover, forest fires) that influence Hg biotransformation (Kelly et al. 2006; Drenner et al. 2013; Klaus et al. 2016; Yang et al. 2016). Recent studies suggest that even fluctuations in weather, such as flooding events, can alter the availability and bioaccumulation of Hg in reptiles. Lázaro et al. (2015) observed that Brazilian caimans (*Caiman yacare*) had higher total mercury (THg) concentrations in their scutes and claws during flood periods than those sampled in drought conditions. Thus, when studying local Hg uptake and effects in biota, researchers must take care to consider all factors that can impact Hg biotransformation within their system.

3 Snakes as Biomarkers of Mercury Exposure

Studies of Hg in wildlife have mostly focused on mammals, fish, birds, and other species of importance to human exposure (e.g., game species or species of immediate economic value). Surprisingly, even though snakes are often important predators co-inhabiting within the same ecosystems, studies of Hg bioaccumulation and its potential impact on snake health are sparse (Drewett et al. 2013). All snakes are either secondary, tertiary, or top predators within their respective ecosystems; thus, they play a crucial role in the transfer of environmental contaminants (Campbell and Campbell 2001).

Snakes exhibit a variety of natural history traits that make them exceptional candidates for ecotoxicological studies. One of the most important traits to consider is their diet. While some snake species are dietary generalists, many are dietary specialists that only consume a specific type of prey (Drewett et al. 2013). For instance, rough green snakes (*Opheodrys aestivus*) and glossy crayfish snakes (*Liodytes [regina] rigida*) specialize on insects and crayfish, respectively (Gibbons and Dorcas 2008; Mason 2008). Diets may also vary among populations within a single species, as has been demonstrated in the viperine watersnake (*Natrix maura*; Lemaire et al. 2018). Some snake species also undergo ontogenetic shifts in dietary preferences, which may impact how snakes at differing life stages accumulate specific contaminants. The brown watersnake (*Nerodia taxispilota*) is a great example of a species whose bioaccumulation potential likely hinges on its ontogenetic dietary shift. Although brown watersnakes are piscivorous throughout their lives, at approximately 60 cm snout-vent length, they shift from a diet that includes a variety of fish prey to a diet mainly comprised of catfish (*Ictalurus* spp.; Mills 2002). Although dietary exposure is the focus of most contaminant studies, dermal contact with contaminated soils may be an underappreciate source of exposure (Weir et al. 2010). Thus, the proportion of time a snake spends buried or in contact with the soil and the propensity of different contaminants to partition in soil layers would both influence exposure and accumulation. Furthermore, because they have relatively small home ranges and high site fidelity (Mills 2002; Beaupre and Douglas 2009), snakes may be particularly valuable biomarkers of local contamination.

Life history traits of snakes also contribute to their potential as bioindicators of Hg pollution. Many species of snakes are relatively long-lived and thus can accumulate high Hg burdens throughout their lifetime. Some large-bodied watersnakes (e.g., brown watersnakes) can live for more than 10 years (Mills 2002). Pit vipers such as the timber rattlesnake (*Crotalus horridus*) are capable of even more impressive lifespans, with maximum estimates of more than 30 years (Brown 2016). Another important life history trait potentially influencing Hg bioaccumulation and exposure routes is reproductive strategy. Snake' reproductive strategies vary but can be broadly categorized as oviparous or viviparous. Furthermore, snakes also exhibit differences in how they invest their energy stores into their offspring. Capital breeders used stored energy, while income breeders use recently acquired energy (Bonnet et al. 1998; Gregory 2006). These variations dictate how contaminants and potentially the extent to which contaminants may be maternally transferred to offspring. For example, animals that are primarily capital breeders may concentrate contaminants in their body reserves, allowing mobilization and circulation of high contaminant levels during vitellogenesis (Meijer and Drent 1999; Rowe 2008).

Our overall understanding of snake ecology is still growing, but more studies are needed to determine what factors may impact their utility as biomarkers of Hg pollution. As mentioned previously, Hopkins (2006) suggested that reptiles may accumulate higher amounts of Hg than other taxa because of their high conversion efficiencies. Furthermore, multiple studies show that snakes can occur at high densities, with substantial biomasses (Houston and Shine 1994; Mills 2002; Willson and Winne 2016). As relatively sedentary, obligate predators occupying a diversity of ecosystems, snakes have the potential to accumulate high amounts of Hg and are likely important contributors to Hg transfer within food webs.

4 Mercury Bioaccumulation in Snakes

As in other species, Hg partitioning in snakes is known to vary among tissue types. Published values in snakes are often opportunistic and typically obtained from destructively sampled individuals. However, in recent years, there has been an increased use of nondestructively sampled tissues for quantifying contaminant burdens to allow for repeated sampling of individuals and to provide the opportunity to investigate the potential sublethal impacts of contaminant exposure in snakes (Hopkins et al. 2001; Jones and Holladay 2006). Thus, it is important for researchers to consider tissue types when interpreting Hg exposure and its effects.

We reviewed the literature (year range 1980–2018) to compile tissue-specific Hg concentration data from snakes (see Table 1). Studies of Hg bioaccumulation in snakes have predominantly focused on taxa (e.g., *Agkistrodon*, *Natrix*, *Nerodia*, *Python*, and *Thamnophis* spp.) that frequent aquatic environments, as these habitats are often associated with high rates of Hg methylation (Scheuhammer et al. 2007). Of the 18 snake species with published studies, only seven species were not in the natricine subfamily of colubrids. *Nerodia* (i.e., watersnakes) were the most commonly sampled genus for Hg exposure (appearing in 12/23 of reviewed articles in Table 1). However, a few studies reported Hg burdens for terrestrial species, including the pine snake (*Pituophis melanoleucus*), the rat snake (*Pantherophis* [Elaphe] *obsoleta*), and the big-eye rat snake (*Ptyas dhumnades*; Burger 1992; Burger et al. 2017; Drewett et al. 2013; Abeysinghe et al. 2017). Our review also highlights a strong geographic bias in the availability of published studies focused on snakes in ecotoxicology, with most of the research occurring in North America (see Table 1).

One of the highest Hg burdens reported in any snake to date was from a northern watersnake (*Nerodia sipedon*) collected in Virginia (tail tip, 13.84 mg/kg dw; Drewett et al. 2013), surpassing values previously reported in snapping turtles (*Chelydra serpentina*) from the same site (Hopkins et al. 2013b). Another study in Taiwan reported that snakes had some of the highest Hg burdens found in biota from Kenting National Park (maximum of 23.9 mg/kg dw Hg); however, they did not report the species of snake sampled in their study. Similarly, aquatic snakes from other study systems have also been documented to have higher Hg burdens than co-occurring top predators (Chumchal et al. 2011; Drewett et al. 2013). In Texas and Louisiana, Chumchal et al. (2011) found that cottonmouths (*A. piscivorus*) attained liver Hg concentrations of 7.46 mg/kg dw – three times that of American alligators (*Alligator mississippiensis*; 2.26 mg/kg dw) from the same site. In the Florida Everglades, invasive Burmese pythons (*Python molurus bivittatus*), which are reported to consume wading birds, mammals, and even alligators (Snow et al. 2007; Dove et al. 2011; Dorcas et al. 2012), had higher muscle Hg concentrations (10.75 mg/kg Hg; Axelrad et al. 2011) than sympatric alligators and fish (Axelrad et al. 2011). Snakes that feed at lower trophic levels but occur near point sources of Hg contamination, however, can still attain high Hg levels, as illustrated by the Hg levels in Virginia northern watersnakes, which were captured near a former acetate fiber production facility (13.84 mg/kg dw; Drewett et al. 2013). Collectively, these studies reveal the propensity for snakes to bioaccumulate high levels of Hg, thereby supporting their utility as biomarkers of Hg contamination.

Table 1 Mean ± SE (followed by range, when provided in corresponding study) mercury (Hg) concentrations (mg/kg) by snake species and tissue type (blood, brain, egg, liver, kidney, muscle, tail tip, whole body)

Species	Mean Hg (mg/kg)	Tissue type	Wet/dry wt	Location	Citation
Agkistrodon piscivorus	0.9 ± 0.1	Muscle	W	Savannah River Site, SC, USA	Burger et al. (2006)
A. piscivorus	0.1 ± 0.04	Blood	W	Savannah River Site, SC, USA	Burger et al. (2006)
A. piscivorus	0.211	Kidney	W	Longhorn Army Ammunitions Plant, TX, USA	Rainwater et al. (2005)[a]
A. piscivorus	0.739	Liver	W	Longhorn Army Ammunitions Plant, TX, USA	Rainwater et al. (2005)[a]
A. piscivorus	0.163	Tail tip	W	Longhorn Army Ammunitions Plant, TX, USA	Rainwater et al. (2005)[a]
A. piscivorus	0.117 ± 0.09	Blood	D	Savannah River Site, SC, USA	Murray et al. (2010)[b]
A. piscivorus	1.204 ± 0.475	Muscle	D	Savannah River Site, SC, USA	Murray et al. (2010)[b]
A. piscivorus	0.145 ± 0.05	Blood	D	Savannah River Site, SC, USA	Murray et al. (2010)[b]
A. piscivorus	1.103 ± 0.167	Muscle	D	Savannah River Site, SC, USA	Murray et al. (2010)[b]
A. piscivorus	0.0135	Blood	W	Old River Slough, TX, USA	Clark et al. (2000)[a]
A. piscivorus	3.292	Muscle	D	Caddo Lake, LA/TX, USA	Chumchal et al. (2011)
A. piscivorus	7.456	Liver	D	Caddo Lake, LA/TX, USA	Chumchal et al. (2011)
Natrix maura	0.145 ± 0.071	Skin	D	Brenne, France	Lemaire et al. (2018)
N. maura	0.289 ± 0.218	Skin	D	Cébron, France	Lemaire et al. (2018)
N. maura	0.439 ± 0.261	Skin	D	Fontenille, France	Lemaire et al. (2018)
N. maura	0.183 ± 0.159	Skin	D	Moëze, France	Lemaire et al. (2018)
N. maura	0.220 ± 0.153	Skin	D	Tour du Valat, France	Lemaire et al. (2018)
N. maura	0.719 ± 0.267	Blood	D	Tour du Valat, France	Lemaire et al. (2018)
N. maura	0.438 ± 0.241	Skin	D	Ons, Spain	Lemaire et al. (2018)
Nerodia spp.	0.18 (0.13–0.21)	Whole body	W	Upper Apalachiola River, FL, USA	Winger et al. (1984)

(continued)

Table 1 (continued)

Species	Mean Hg (mg/kg)	Tissue type	Wet/dry wt	Location	Citation
Nerodia spp.	0.29 (0.17–0.38)	Whole body	W	Lower Apalachiola River, FL, USA	Winger et al. (1984)
Nerodia fasciata	0.6 ± 0.05	Muscle	W	Savannah River Site, SC, USA	Burger et al. (2006)
N. fasciata	0.4 ± 0.05	Blood	W	Savannah River Site, SC, USA	Burger et al. (2006)
N. fasciata	0.4 ± 0.047	Blood	W	Savannah River Site, SC, USA	Burger et al. (2007)
N. fasciata	0.192 ± 0.014	Tail tip	W	Savannah River Site, SC, USA	Burger et al. (2007)
N. fasciata	1.857 ± 0.452	Liver	W	Savannah River Site, SC, USA	Burger et al. (2007)
N. fasciata	0.379 ± 0.057	Blood	D	Savannah River Site, SC, USA	Murray et al. (2010)[b]
N. fasciata	0.538 ± 0.047	Muscle	D	Savannah River Site, SC, USA	Murray et al. (2010)[b]
N. fasciata	0.460 ± 0.084	Blood	D	Savannah River Site, SC, USA	Murray et al. (2010)[b]
N. fasciata	0.860 ± 0.081	Muscle	D	Savannah River Site, SC, USA	Murray et al. (2010)[b]
Nerodia floridana	0.327 ± 0.028	Tail tip	D	Savannah River Site, SC, USA	Russell et al. (2016)
Nerodia rhombifer	0.0613	Blood	W	Private Lake, TX, USA	Clark et al. (2000)[a]
N. rhombifer	0.146	Blood	W	Old River Slough, TX, USA	Clark et al. (2000)[a]
Nerodia sipedon	0.061 ± 0.001	Egg	W	Walland, TN, USA	Burger et al. (2005)
N. sipedon	0.289 ± 0.133	Testes	W	Walland, TN, USA	Burger et al. (2005)
N. sipedon	0.423 ± 0.028	Skin	W	Walland, TN, USA	Burger et al. (2005)
N. sipedon	1.121 ± 0.173 (0.209–3.505)	Kidney	W	Oak Ridge, TN, USA	Campbell et al. (2005)
N. sipedon	1.403 ± 0.214 (0.220–3.795)	Liver	W	Oak Ridge, TN, USA	Campbell et al. (2005)
N. sipedon	0.582 ± 0.047 (0.051–1.015)	Muscle	W	Oak Ridge, TN, USA	Campbell et al. (2005)
N. sipedon	0.372 ± 0.0461 (0.141–0.816)	Blood	W	Oak Ridge, TN, USA	Campbell et al. (2005)

N. sipedon	0.382 ± 0.032 (0.042–0.784)	Kidney	W	Walland, TN, USA	Campbell et al. (2005)
N. sipedon	0.75 ± 0.076 (0.090–1.161)	Liver	W	Walland, TN, USA	Campbell et al. (2005)
N. sipedon	0.741 ± 0.049 (0.224–1.630)	Muscle	W	Walland, TN, USA	Campbell et al. (2005)
N. sipedon	0.436 ± 0.064 (0.009–1.420)	Blood	W	Walland, TN, USA	Campbell et al. (2005)
N. sipedon	0.45	Whole body	W	Lake Michigan, MI, USA	Heinz et al. (1980)
N. sipedon	0.128 ± 0.026	Blood	W	Raritan Canal, NJ, USA	Burger et al. (2007)
N. sipedon	0.136 ± 0.039	Kidney	W	Raritan Canal, NJ, USA	Burger et al. (2007)
N. sipedon	0.303 ± 0.091	Liver	W	Raritan Canal, NJ, USA	Burger et al. (2007)
N. sipedon	0.357 ± 0.049	Muscle	W	Raritan Canal, NJ, USA	Burger et al. (2007)
N. sipedon	0.159 ± 0.023	Skin	W	Raritan Canal, NJ, USA	Burger et al. (2007)
N. sipedon	0.417 ± 0.042	Blood	W	Oak Ridge, TN, USA	Burger et al. (2007)
N. sipedon	0.671 ± 0.038	Muscle	W	Oak Ridge, TN, USA	Burger et al. (2007)
N. sipedon	1.024 ± 0.115	Liver	W	Oak Ridge, TN, USA	Burger et al. (2007)
N. sipedon	0.29 ± 0.01 (0.23–0.37)	Tail tip	D	Middle River, VA, USA	Drewett et al. (2013)
N. sipedon	0.49 ± 0.07 (0.16–0.92)	Tail tip	D	South River, VA, USA	Drewett et al. (2013)
N. sipedon	4.85 ± 0.29 (2.25–13.84)	Tail tip	D	South River, VA, USA	Drewett et al. (2013)
N. sipedon	2.24 ± 0.42 (0.03–7.04)	Blood	W	Virginia, USA	Drewett et al. (2013)
N. sipedon	0.2 ± 0.11 (0.06–1.09)	Whole body	D	Middle River, VA, USA	Chin et al. (2013b)
N. sipedon	3.42 ± 0.45 (1.08–10.10)	Whole body	D	South River, VA, USA	Chin et al. (2013b)

(continued)

Table 1 (continued)

Species	Mean Hg (mg/kg)	Tissue type	Wet/dry wt	Location	Citation
N. sipedon	0.0037 ± 0.0001	Fecal	W	MTSU, TN, USA	Cusaac et al. (2016)[c]
N. sipedon	0.0070 ± 0.001	Fecal	W	MTSU, TN, USA	Cusaac et al. (2016)[c]
N. sipedon	2.87 ± 1.53	Fecal	W	MTSU, TN, USA	Cusaac et al. (2016)[c]
N. sipedon	0.112 ± 0.019	Muscle	W	MTSU, TN, USA	Cusaac et al. (2016)[c]
N. sipedon	0.199 ± 0.047	Muscle	W	MTSU, TN, USA	Cusaac et al. (2016)[c]
N. sipedon	13.2 ± 2.58	Muscle	W	MTSU, TN, USA	Cusaac et al. (2016)[c]
N. sipedon	0.116 ± 0.023	Liver	W	MTSU, TN, USA	Cusaac et al. (2016)[c]
N. sipedon	0.0935 ± 0.021	Liver	W	MTSU, TN, USA	Cusaac et al. (2016)[c]
N. sipedon	13.5 ± 8.16	Liver	W	MTSU, TN, USA	Cusaac et al. (2016)[c]
N. sipedon	0.286 ± 0.059	Skin	W	MTSU, TN, USA	Cusaac et al. (2016)[c]
N. sipedon	0.375 ± 0.032	Skin	W	MTSU, TN, USA	Cusaac et al. (2016)[c]
N. sipedon	95.4 ± 21.1	Skin	W	MTSU, TN, USA	Cusaac et al. (2016)[c]
Nerodia taxispilota	0.7 ± 0.1	Muscle	W	Savannah River Site, SC, USA	Burger et al. (2006)
N. taxispilota	0.7 ± 0.15	Blood	W	Savannah River Site, SC, USA	Burger et al. (2006)
N. taxispilota	0.611 ± 0.180	Blood	D	Savannah River Site, SC, USA	Murray et al. (2010)[b]
N. taxispilota	0.644 ± 0.100	Muscle	D	Savannah River Site, SC, USA	Murray et al. (2010)[b]
N. taxispilota	0.923 ± 0.005	Blood	D	Savannah River Site, SC, USA	Murray et al. (2010)[b]
N. taxispilota	0.971 ± 0.179	Muscle	D	Savannah River Site, SC, USA	Murray et al. (2010)[b]
Pantherophis (Elaphe) guttata	0.403 ± 0.089	Skin	D	Virginia Tech, VA, USA	Jones and Holladay (2006)[c]
Pantherophis (Elaphe) obsoleta	0.26 ± 0.09 (0.05–0.89)	Tail tip	D	South River, VA, USA	Drewett et al. (2013)
Pituophis melanoleucus	0.13 ± 0.027	Whole body	D	New Jersey, USA	Burger (1992)[a]
P. melanoleucus	0.28 ± 0.047	Skin	D	New Jersey, USA	Burger (1992)[a]

P. melanoleucus	0.46 ± 0.078	Liver	W	New Jersey, USA	Burger et al. (2017)
P. melanoleucus	0.12 ± 0.035	Kidney	W	New Jersey, USA	Burger et al. (2017)
P. melanoleucus	0.76 ± 0.012	Muscle	W	New Jersey, USA	Burger et al. (2017)
P. melanoleucus	0.42 ± 0.007	Skin	W	New Jersey, USA	Burger et al. (2017)
P. melanoleucus	0.41 ± 0.009	Heart	W	New Jersey, USA	Burger et al. (2017)
P. melanoleucus	0.27 ± 0.005	Blood	W	New Jersey, USA	Burger et al. (2017)
Ptyas dhumnades	9.77 ± 0.68	Tail tip	D	Guizhou, China	Abeysinghe et al. (2017)
P. dhumnades	2.36 ± 0.175	Tail tip	D	Guizhou, China	Abeysinghe et al. (2017)
Python molurus bivittatus	3.6 (0.14–10.75)	Muscle	W	Everglades, FL, USA	Axelrad et al. (2011)
Liodytes (Regina) septemvittata	4.59 ± 0.38 (1.90–6.00)	Tail tip	D	South River, VA, USA	Drewett et al. (2013)
Thamnophis gigas	0.571 ± 0.108 (0.08–1.64)	Liver	W	Sacramento Valley, CA, USA	Wylie et al. (2009)
T. gigas	0.077 ± 0.010 (0.01–0.18)	Brain	W	Sacramento Valley, CA, USA	Wylie et al. (2009)
T. gigas	0.083 ± 0.019 (0.02–0.32)	Tail tip	W	Sacramento Valley, CA, USA	Wylie et al. (2009)
Thamnophis sauritus	0.58 ± 0.12	Whole body	D	Mobile-Tensaw River, AL, USA	Albrecht et al. (2007)
Thamnophis sirtalis	0.303	Whole body	W	Lake Michigan Spider Island, USA	Heinz et al. (1980)
T. sirtalis	1.28 ± 0.32 (0.08–2.53)	Tail tip	D	South River, VA, USA	Drewett et al. (2013)
Unknown spp.	5.41 ± 2.28 (0.16–23.9)	Muscle	D	Kenting National Park, Taiwan	Hsu et al. (2006)

Table is based on 23 published studies in the literature (year range 1980–2018). We performed a literature search for publications that focused on snakes, mercury, ecological risk assessments, and biomarkers. We searched for these papers in multiple databases including Web of Science and Google Scholar. We also used references cited in publications from our literature search to thoroughly search for relevant studies

[a]Geometric means

[b]Dry vs. wet weight not explicitly listed

[c]Controlled exposure experiment

5 Effects of Mercury on Snake Health and Immune Status

5.1 Mercury Effects in Wildlife

The effects of Hg on wildlife vary widely, but in birds and mammals, these effects are generally characterized by aberrations in the endocrine, immune, neurological, and reproductive systems (Spalding et al. 2000; Tan et al. 2009; Fallacara et al. 2011). Many of the organ systems are similar across taxa. Thus, the biological effects of Hg exposure in reptiles may have similar consequences as reported for other taxa. It is important to note, however, that both exposure and sensitivity to contaminants can vary among taxa in the same study system (Weir et al. 2010). Unfortunately, relatively little is known about the toxicological significance of Hg exposure in snakes and other reptiles compared to birds, fish, and mammals. The few studies to date on the effects of Hg on reptiles, especially snakes, suggest that they may be more resilient to contaminants relative to other taxa (Wolfe et al. 1998; Bazar et al. 2002; Chin et al. 2013b). However, given the propensity of snakes to accumulate high levels of Hg coupled with the sublethal effects observed in other species, it is quite likely that snakes could be at risk for compromised health.

5.2 Maternally Transferred Mercury in Snakes

Maternal transfer of contaminants in wildlife is an important route of exposure to consider in ecotoxicological research. There are only a handful of studies that examined maternal transfer of contaminants in snakes, and the majority of them focused on Hg (Hopkins et al. 2004; Chin et al. 2013a, b; Cusaac et al. 2016). Chin et al. (2013a) found that high levels of maternally transferred Hg did not significantly impact maternal reproductive output or embryonic survival in northern watersnakes collected in Virginia. Neonates from the same study system were then subjected to tests that gauged their foraging, learning, and locomotor abilities (Chin et al. 2013b). They found that food motivation and striking efficiency in neonates were negatively correlated with Hg burdens. If these behavioral deficits translate to a wild setting, they could lead to reduced growth and fitness in neonates produced by highly contaminated mothers.

In a more recent study involving an artificial maternal Hg transfer technique in which female northern watersnakes were force-fed pills with MeHg during pregnancy, neither corticosterone (CORT) levels nor white blood cell counts in offspring were affected when compared to control offspring (Cusaac et al. 2016). However, absolute baseline level for CORT could not be obtained, as evidenced by control neonates also having maternally transferred Hg even though their mothers were not exposed to MeHg during the experimental trial. The most noteworthy observation from this study was that three (3/17) Hg-exposed mothers died and all three were in either the low- (0.1 mg/kg) or high-dose (10 mg/kg) MeHg groups. Furthermore, the single female mortality from the high-dose group presented with symptoms that were consistent with acute Hg exposure (e.g., lethargy, lack of coordination). Overall, the little information that exists for Hg effects in snakes suggests that northern watersnakes, and perhaps other snake species, may be more tolerant of Hg exposure compared to other taxa, but more research is needed (Chin et al. 2013a, b; Cusaac et al. 2016).

5.3 Mercury and Reptile Immunotoxicology

The vertebrate immune response is sensitive to Hg exposure, as many studies in mammals and birds show that Hg may negatively affect cell proliferation and regulation of cytokines and chemokines and potentially cause cell death (Lewis et al. 2013; Desforges et al. 2016; Gardner and Nyland 2016). Studies examining the effects of contaminants on the reptilian immune system are lacking. In fact, most literature reviews in wildlife immunotoxicology do not include sections for reptiles due to a paucity of research in this field (Keller et al. 2006). Of the two reported studies that examined Hg's impact on reptilian immunity, one study found that leukocyte counts in wild loggerhead sea turtles (*Caretta caretta*) were negatively correlated with blood Hg, suggesting that Hg exposure caused measurable immunosuppression (Day et al. 2007). Yet, in the other study, northern watersnakes collected from a site contaminated with Hg exhibited no differences in wound healing, an indirect measure of innate immunity, compared to snakes from a reference site (Hopkins et al. 2013a). There are numerous host and environmental factors that likely impacted the outcomes of these two studies. Still, based on studies in other species, it is possible that Hg exposure can adversely modulate the snake immune system leading to higher rates of disease and other health issues (Scheuhammer et al. 2007). This is a relevant concern as some snake populations are currently under threat due to emerging diseases, such as snake fungal disease (Lorch et al. 2016). Exposure to additional stressors such as contaminants that disrupt the snake immune system could increase susceptibility to infection or disease. Further research is needed to elucidate the relationship between contaminants, immunity, and health and to better understand their potential individual- and population-level consequences in snakes.

6 Snakes and Mercury Transfer

6.1 Snakes and Mercury Transfer in Aquatic and Terrestrial Ecosystems

An increasingly important topic in ecotoxicological studies is the role of species in linking aquatic and terrestrial food webs (Cristol et al. 2008; Sullivan and Rodewald

2012; Leaphart 2017). Snakes facilitate the movement of energy and contaminants across ecosystems in their roles as both predators and prey. For example, banded watersnakes (*Nerodia fasciata*) and black swamp snakes (*Liodytes [Seminatrix] pygaea*) consume large quantities of terrestrial amphibians returning to wetlands to breed (Willson and Winne 2016), resulting in large transfers of energy to the aquatic habitat from the surrounding terrestrial environment ($>$150,000 kJ ha^{-1} annually in an 10-ha isolated wetland in South Carolina, USA).

Watersnakes (*Nerodia* spp.) can also play an important role in contaminant transfer from the aquatic to the terrestrial environment. In addition to being subject to predation by aquatic species such as fish, watersnakes are also consumed by terrestrial predators, including other snakes (e.g., coachwhips, king snakes, and racers), mammals (e.g., raccoons and armadillos), and birds (e.g., hawks and owls; Mushinsky and Miller 1993; Voris and Murphy 2002; Gibbons and Dorcas 2004; Willson and Winne 2016). Thus, due to their small size, neonates in particular may be prone to predation. The brown watersnake and the diamondback watersnake (*Nerodia rhombifer*) are large-bodied, piscivorous watersnakes that inhabit rivers and permanent bodies of water in the USA (Mills 2004; Keck 2004). Their strictly piscivorous diet likely puts them at risk for bioaccumulation of high amounts of MeHg. In addition, these species are known to reproduce annually, with the largest females capable of producing upward of 60 neonates (Mills 2004; Keck 2004). If females maternally transfer Hg to their young, as reported in northern watersnakes (Chin et al. 2013b), these neonates may potentially spread large amounts of Hg from aquatic sources to terrestrial predators.

6.2 Snakes and Risk of Human Exposure to Contaminants

Another often overlooked hazard of contaminants in snakes is the risk of human exposure. Human consumption of snakes may be uncommon in the USA, but in other countries, snakes are commonly collected for medicine or food and are even considered a delicacy (Klemens and Thorbjarnarson 1995; Schneider et al. 2013). In Cambodia, some records show that during the monsoon season, upward of 8,500 watersnakes can be collected per day for feeding alligators and for human consumption. In addition, estimates of snake consumption by humans in Svay Rieng, Cambodia, are approximately 0.19 kg/person/year (Hortle 2007). Although snakes are not commonly consumed in the USA, reports of Florida residents eating "Everglades pizza" have worried state officials because this dish often includes American alligator, frog legs, and invasive Burmese python (Snyder 2012), which accumulate high amounts of Hg, thereby putting humans at risk (Axelrad et al. 2011).

7 Conclusions

Despite being one of the most well-studied contaminants around the globe, relatively little is known about Hg bioaccumulation in snakes. Snakes are middle to upper trophic level predators within ecosystems; thus, understanding their role in Hg transfer is crucial when performing accurate risk assessments. Studies show that snakes that inhabit aquatic environments (e.g., *Nerodia*, *Python*, *Agkistrodon* spp.) can accumulate markedly high amounts of Hg relative to other taxa in the same system. Because viviparous watersnake species can also maternally transfer Hg to their offspring, it is also important to consider the multiple routes by which snakes can transfer and mobilize Hg within an ecosystem. Furthermore, countries with human populations that rely on snakes as a food source should seriously consider Hg bioaccumulation in snakes as a likely avenue for human exposure, particularly in areas where environmental Hg levels are known to be elevated.

Little is known about the impact of acute or chronic Hg exposure on snakes and their health (i.e., immunology, metabolism, and overall physiology). The limited studies available suggest at least some species (e.g., northern watersnakes) may be tolerant to high levels of Hg (Chin et al. 2013a, b; Cusaac et al. 2016). If widely applicable, their resilience bodes well for the snakes, but the implications are that snakes may readily transport high amounts of Hg and yet not show signs of clinical illness. However, it is also important to note that many aspects of snake health remain poorly known or even unexplored in the scientific literature (e.g., immunotoxicity). Thus, further research is needed to clarify the relationships between Hg body burdens and snake health.

8 Summary

In this review, we emphasize the utility of snakes as important biomarkers of Hg exposure and as critical links for Hg transfer in the environment. We also sought to stress the lack of studies that focus on Hg bioaccumulation and concomitant effects in snakes. Though disdain for snakes persists throughout much of society, snakes are important to biodiversity and overall ecosystem health. Snakes are facing threats from a variety of sources including pollution, emerging diseases, invasive species, and habitat destruction. Many snake species are listed as threatened or endangered, and their ultimate persistence will rely on a more comprehensive understanding of the potential impacts of widespread contaminants on their health.

Acknowledgments Preparation of this manuscript was supported by an assistantship through the University of Georgia's Interdisciplinary Toxicology Program and the Savannah River Ecology Laboratory, as well as the Department of Energy under award number DE-EM0004391 to the University of Georgia Research Foundation and by the Savannah River Nuclear Solutions – Area Completions Project.

References

Abeysinghe KS, Qiu G, Goodale E, Anderson CWN, Bishop K, Evers DC, Goodale MW, Hintelmann H, Liu S, Mammides C, Quan R, Wang J, Wu P, Xu X, Yang X, Feng X (2017) Mercury flow through an Asian rice-based food web. Environ Pollut 229:219–228

Albrecht J, Abalos M, Rice TM (2007) Heavy metal levels in ribbon snakes (*Thamnophis sauritus*) and anuran larvae from the Mobile-Tensaw River Delta, Alabama, USA. Arch Environ Contam Toxicol 53(4):647–654

Axelrad DM, Lange T, Gabriel MC, Aiken GR, Brandon A, Cunningham MW, DeBusk T, Dierberg F, Donner BA, Frederick P, Gilmour C, Harris R, Jansen D, Krabbenhoft DP, McCray JM, Orem WH, Oronato DP, Pollman DC, Rumbold DG, White G, Wright AL, Ye R (2011) Mercury and sulfur monitoring, research and environmental assessment in South Florida. In: Graham SH, McCallum KK (eds) South Florida environmental report. West Palm Beach, pp 3B1–3B53

Bazar M, Holtzman D, Adair B, Gresens S (2002) Effects of dietary methylmercury in juvenile corn snakes (*Elaphe guttata*). In: SETAC 23rd annual meeting

Beaupre SJ, Douglas LE (2009) Snakes as indicators and monitors of ecosystem properties. In: Mullins SJ, Seigel RA (eds) Snakes: ecology and conservation. Cornell University Press, Ithaca, pp 244–261

Boening DW (2000) Ecological effects, transport, and fate of mercury: a review. Chemosphere 40:1335–1351

Bonnet X, Bradshaw D, Shine R (1998) Capital versus income breeding: an ectothermic perspective. Oikos 83:333–342

Brown WS (2016) Lifetime reproduction in a northern metapopulation of timber rattlesnakes (*Crotalus horridus*). Herpetologica 72(4):331–342

Burger JS (1992) Trace element levels in pine snake hatchlings: tissue and temporal differences. Arch Environ Contam Toxicol 22:209–213

Burger JS, Campbell KR, Campbell TS, Shukla T, Jeitner C, Gochfeld M (2005) Use of skin and blood as nonlethal indicators of heavy metal contamination in northern watersnakes (*Nerodia sipedon*). *Arch Environ Contam Toxicol* 49(2):232–238

Burger JS, Murray S, Gaines KF, Novak JM, Punshon T, Dixon C, Gochfield M (2006) Elemental levels in snakes in South Carolina: differences between a control site and exposed site on the Savannah River Site. Environ Monit Assess 112:35–52

Burger JS, Campbell K, Murray S, Campbell TS, Gaines KF, Jeitner C, Shukla T, Burke S, Gochfield M (2007) Metal levels in blood, muscle and liver of watersnakes (*Nerodia* spp.) from New Jersey, Tennessee and South Carolina. Sci Total Environ 373:556–563

Burger JS, Gochfield M, Jeitner C, Zappalorti R, Pittfield T, DeVito E (2017) Arsenic, cadmium, chromium, lead, mercury, and selenium concentrations in pine snakes (*Pituophis melanoleucus*) from the New Jersey pine barrens. Arch Environ Contam Toxicol 72:586–595

Campbell KR, Campbell TS (2001) The accumulation and effects of environmental contaminants on snakes: a review. Environ Monit Assess 70:253–301

Campbell KR, Campbell TS, Burger JS (2005) Heavy metal concentrations in northern watersnakes (Nerodia sipedon) from east fork Poplar Creek and the Little River, East Tennessee, USA. Arch Environ Contam Toxicol 49:239

Chen L, Wang HH, Liu JF, Tong YD, Ou LB, Zhang W, Hu D, Chen C, Xang XJ (2014) Intercontinental transport and deposition patterns of atmospheric mercury from anthropogenic emissions. Atmos Chem Phys 14:10163–10176

Chin SY, Willson JD, Cristol DA, Drewett DVV, Hopkins WA (2013a) Altered behavior of neonatal watersnakes (*Nerodia sipedon*) exposed to maternally transferred mercury. Environ Pollut 176:144–150

Chin SY, Willson JD, Cristol DA, Drewett DVV, Hopkins WA (2013b) High levels of maternally transferred mercury do not affect reproductive output or embryonic survival of northern watersnakes (*Nerodia sipedon*). Environ Toxicol Chem 32(3):619–626

Chumchal MM, Rainwater TR, Osborn SC, Roberts AP, Abel MT, Cobb GP, Smith PN, Bailey FC (2011) Mercury speciation and biomagnifications in the food web of Caddo Lake (Texas/ Louisiana, USA), a subtropical freshwater ecosystem. Enrivon Toxicol Chem 30:1153–1162

Clark DR Jr, Bickham JW, Baker DL, Cowman DF (2000) Environmental contaminants in Texas, USA, wetland reptiles: evaluation using blood samples. Environ Toxicol Chem 19 (9):2259–2265

Cristol DA, Brasso RL, Condon AM et al (2008) The movement of aquatic mercury through terrestrial food webs. Science 320:335

Cusaac JPW, Kremer V, Wright R, Henry C, Otter RR, Bailey FC (2016) Effects of maternally transferred methylmercury on stress physiology in northern watersnake (*Nerodia sipedon*) neonates. Bull Environ Contam Toxicol 96:725–731

Day RD, Segars AL, Arendt MD, Lee AM, Peden-Adams MM (2007) Relationship of blood mercury levels to health parameters in the loggerhead sea turtle (*Caretta caretta*). Environ Health Perspect 115(10):1421–1428

Desforges JPW, Sonne C, Levin M et al (2016) Immunotoxic effects of environmental pollutants in marine mammals. Environ Int 86:126–139

Dorcas ME, Willson JD, Reed RN, Snow RW, Rochford MR, Miller MA, Meshaka WE Jr, Andreadis PT, Mazzoti FJ, Romagosa CM, Hart KM (2012) Severe mammal declines coincide with proliferation of invasive Burmese pythons in Everglades National Park. PNAS 109 (7):2418–2422

Dove CJ, Snow RW, Rochford MR, Mazzotti FJ (2011) Birds consumed by the invasive Burmese python (*Python molurus bivittatus*) in Everglades National Park, Florida, USA. Wilson J Ornithol 123(1):126–131

Drenner RW, Chumchal MM, Jones CM, Lehmann CMB, Gay DA, Donato DI (2013) Effects of mercury deposition and coniferous forest on the mercury contamination of fish in the South Central United States. Environ Sci Technol 47:1274–1279

Drewett DVV, Willson JD, Cristol DA, Chin SY, Hopkins WA (2013) Inter-and intraspecific variation in mercury bioaccumulation by snakes inhabiting a contaminated river floodplain. Environ Toxicol Chem 32:1178–1186

Durso AM, Willson JD, Winne CT (2011) Needles in haystacks: estimating detection probability and occupancy of rare and cryptic snakes. Biol Conserv 144:1508–1515

Fallacara DM, Halbrook RS, French JB (2011) Toxic effects of dietary methylmercury on immune function and hematology in American Kestrels (*Falco sparverius*). Environ Toxicol Chem 30 (6):1320–1327

Gardner RM, Nyland JF (2016) Immunotoxic effects of mercury. In: Esser C (ed) Environmental influences on the immune system. Springer, Vienna, pp 273–302

Gibbons JW, Dorcas ME (2004) North American Watersnakes: a natural history. University of Oklahoma Press, Norman

Gibbons JW, Dorcas ME (2008) Glossy crayfish snake (*Regina rigida*). In: Jensen JB, Camp CD, Gibbons JW, Elliot MJ (eds) Amphibians and reptiles of Georgia. University of Georgia Press, Athens, pp 393–394, 575 pp

Gibbons JW, Scott DW, Ryan TJ, Buhlmann KA, Tuberville TD, Metts BS, Greene JL, Mills T, Leiden M, Poppy S, Winne CT (2000) The global decline of reptiles, Déjà vu amphibians. Bioscience 50:653–666

Gregory PT (2006) Influence of income and capital on reproduction in a viviparous snake: direct and indirect effects. J Zool 270(3):414–419

Grillitsch B, Schiesari L (2010) The ecotoxicology of metals in reptiles. In: Sparling DW, Linder G, Bishop AC, Krest S (eds) Ecotoxicology of amphibians and reptiles. Society of Environmental Toxicology and Chemistry (SETAC), Pensacola, pp 337–448

Heinz GH, Haseltine SD, Hall RJ, Krynitsky AJ (1980) Organochlorine and mercury residues in snakes from Pilot and Spider Islands, Lake Michigan – 1978. Bull Environ Contam Toxicol 25 (1):738–743

Hopkins WA (2000) Reptile toxicology: challenges and opportunities on the last frontier in vertebrate ecotoxicology. Environ Toxicol Chem 19:2391–2393

Hopkins WA (2006) Use of tissue residues in reptile ecotoxicology: a call for integration and experimentalism. In: Gardner SC, Oberdörster E (eds) Toxicology of reptiles. CRC Press, Boca Raton, pp 35–62

Hopkins WA, Roe JD, Snodgrass JW et al (2001) Nondestructive indices of trace element exposure in squamate reptiles. Environ Pollut 115:1–7

Hopkins WA, Staub BP, Baionno JA et al (2004) Trophic and maternal transfer of selenium in brown house snakes (*Lamprophis fuliginosus*). Ecotoxicol Environ Saf 58:285–293

Hopkins BC, Chin SY, Willson JD, Hopkins WA (2013a) Like mother, like offspring: maternal and offspring wound healing correlate in snakes. J Exp Biol 216:2545–2547

Hopkins BC, Willson JD, Hopkins WA (2013b) Mercury exposure is associated with negative effects on turtle reproduction. Environ Sci Technol 47:2416–2422

Hortle KG (2007) Consumption and yield of fish and other aquatic animals from the lower Mekong Basin. MRC technical paper 16, pp 1–88

Houston D, Shine R (1994) Population demography of Arafura filesnakes (Serpentes: Acrochordidae) in tropical Australia. J Herpetol 28:273–280

Hsu MJ, Selvaraj K, Agoramoorthy G (2006) Taiwan's industrial heavy metal pollution threatens terrestrial biota. Environ Pollut 143:327–334

IUCN (2017) The IUCN red list of threatened species. Version 2016-3. http://www.iucnredlist.org. Accessed 12 Mar 2017

Jones DE, Holladay SD (2006) Excretion of three heavy metal in the shed skin of exposed corn snakes (*Elaphe guttata*). Ecotoxicol Environ Saf 64:221–225

Keck MB (2004) Species account of the diamondback watersnake (*Nerodia rhombifer*). In: Gibbons JW, Dorcas ME (eds) North American Watersnakes: a natural history. University of Oklahoma Press, Norman, pp 138–177

Keller JM, Peden-Adams M, Aguirre A (2006) Immunotoxicity and implications for reptilian health. In: Gardner SC, Oberdörster E (eds) Toxicology of reptiles. CRC Press, Boca Raton, pp 199–240

Kelly EN, Schindler DW, St. Louis VL et al (2006) Forest fire increases mercury accumulation by fishes via food web restructuring and increased mercury inputs. PNAS 103(51):19380–19385

Klaus JE, Hammerschmidt CR, Costello DM, Burton GA (2016) Net methylmercury production in 2 contrasting stream sediments and associated accumulation and toxicity to periphyton. Environ Toxicol Chem 35(7):1759–1765

Klemens MW, Thorbjarnarson JB (1995) Reptiles as a food resource. Biodivers Conserv 4:281–298

Lamborg CH, Hammerschmidt CR, Bowman KL et al (2014) A global ocean inventory of anthropogenic mercury based on water column measurements. Nature 512:65–68

Lázaro WL, Oliveira RF, Santos-Filho M et al (2015) Non-lethal sampling for mercury evaluation in crocodilians. Chemosphere 138:25–32

Leaphart JC (2017) Environmental fate of radiocesium in aquatic and semi-aquatic biota on the U.S. Department of Energy's Savannah River Site. MS thesis, University of Georgia, Athens, 105 pp

Lemaire J, Bustamante P, Oliver A et al (2018) Determinants of mercury contamination in viperine snakes, *Natrix maura*, in Western Europe. Sci Total Environ 635:20–25

Lewis CA, Cristol DA, Swaddle JP et al (2013) Decreased immune response in zebra finches exposed to sublethal doses of mercury. Arch Environ Contam Toxicol 64:327–336

Lorch JM, Knowles S, Lankton JS et al (2016) Snake fungal disease: an emerging threat to wild snakes. Philos Trans R Soc Lond Ser B Biol Sci 371(1709):20150457

Mason PA (2008) Rough green snake (*Opheodrys aestivus*). In: Jensen JB, Camp CD, Gibbons JW, Elliot MJ (eds) Amphibians and reptiles of Georgia. University of Georgia Press, Athens, pp 386–387, 575 pp

Meijer T, Drent R (1999) Re-examination of the capital and income dichotomy in breeding birds. Ibis 141:399–414

Mills MS (2002) Ecology and life history of the brown watersnake (*Nerodia taxispilota*). PhD dissertation, University of Georgia, Georgia

Mills MS (2004) Species account of the brown watersnake (*Nerodia taxispilota*). In: Gibbons JW, Dorcas ME (eds) North American Watersnakes: a natural history. University of Oklahoma Press, Norman, pp 196–211
Murray SM, Gaines KF, Novak JM, Gochfield M, Burger JS (2010) DNA double-strand breakage as an endpoint to examine metal and radionuclide exposure effects to watersnakes on a nuclear industrial site. Hum Ecol Risk Assess 16:282–300
Mushinsky HR, Miller DE (1993) Predation on watersnakes: ontogenetic and interspecific considerations. Copeia 1993(3):660–665
Newman MC (2015) Risk assessment of contaminants. In: Fundamentals of ecotoxicology, 4th edn. CRC Press, Boca Raton, pp 397–423
Pough FH, Andrews RM, Cadle JE et al (1998) In: Pough FH (ed) Herpetology. Prentice-Hall, Upper Saddle River, pp 1–577
Rainwater TR, Reynolds KD, Cañas JE, Cobb GP, Anderson TA, McMurry ST, Smith PN (2005) Organochlorine pesticides and mercury in cottonmouths (*Agkistrodon piscivorus*) from northeastern Texas, USA. Environ Toxicol Chem 24(3):665–673
Reading CJ, Luiselli LM, Akani GC et al (2010) Are snake populations in widespread decline? Biol Lett 6(6):777–780
Rowe CR (2008) "The calamity of so long life": life histories, contaminants, and potential emerging threats to long-lived vertebrates. Bioscience 58(7):623–631
Russell AL, Brown MK, Lambert MM, Tuberville TD, Pilgrim MA (2016) Mercury bioaccumulation in Florida green watersnakes (*Nerodia floridana*) inhabiting former nuclear cooling reservoirs on the Savannah River Site. USC Upstate Student Res J X(Fall):17–25
Scheuhammer AM, Meyer MW, Sandheinrich MB, Murray MW (2007) Effects of environmental methylmercury on the health of wild birds, mammals, and fish. Ambio 36(1):12–18
Schneider L, Maher W, Green A, Vogt RC (2013) Mercury contamination in reptiles: an emerging problem with consequences for wild life and human health. In: Kim K, Brown RJC (eds) Mercury: sources, applications and health impacts. Nova Science Publishers, New York, pp 173–232
Selin NE (2009) Global biogeochemical cycling of mercury: a review. Annu Rev Environ Resour 34:43–63
Shine R, Bonnet X (2000) Snakes: a new 'model organism' in ecological research? Trends Ecol Evol 15(6):221–222
Snow RW, Brien ML, Cherkiss S et al (2007) Dietary habits of Burmese python, *Python molurus bivittatus*, from Everglades National Park, Florida. Herpetol Bull 101:5–7
Snyder DD (2012) Florida officials: don't eat Python meat. https://www.outsideonline.com/1794941/florida-officals-dont-eat-python-meat. Accessed 10 Mar 2017
Spalding MG, Frederick PC, McGill HC et al (2000) Histologic, neurologic, and immunologic effects of methylmercury in captive great egrets. J Wildl Dis 36(3):423–435
Sullivan SMP, Rodewald AD (2012) In a state of flux: the energetic pathways that move contaminants from aquatic to terrestrial environments. Environ Toxicol Chem 31(6):1–9
Tan SW, Meiller JC, Mahaffey KR (2009) The endocrine effects of mercury in humans and wildlife. Crit Rev Toxicol 39(3):228–269
Tweedy BN, Drenner RW, Chumchal MM, Kennedy JH (2013) Effects of fish on emergent insect-mediated flux of methylmercury across a gradient of contamination. Environ Sci Technol 47:1614–1619
Vitt LJ, Caldwell JP (2014) In: Gomez K, Gonzalez P (eds) Herpetology: an introductory biology of amphibians and reptiles. Elsevier, Amsterdam, pp 1–626
Voris HK, Murphy JC (2002) The prey and predators of Homalopsine snakes. J Nat Hist 36:1621–1632
Weir SM, Suski JG, Salice CJ (2010) Ecological risk of anthropogenic pollutants to reptiles: evaluating assumptions of sensitivity and exposure. Environ Pollut 158(12):3596–3606
Willson JD, Winne CT (2016) Evaluating the functional importance of secretive species: a case study of aquatic snake predators in isolated wetlands. J Zool 298:266–273

Willson JD, Winne CT, Todd BD (2011) Ecological and methodological factors affecting detectability and population evaluation in elusive species. J Wildl Manag 75:36–45

Winger PV, Sieckman C, May TW, Johnson WW (1984) Residues of organochlorine insecticides polychlorinated biphenyls, and heavy metals in biota from Apalachicola River, Florida, 1978. J Assoc Off Anal Chem 67(2):325–333

Wolfe MF, Schwarzbach S, Sulaiman RA (1998) Effects of mercury on wildlife: A comprehensive review. *Environ Toxicol Chem* 17(2):146–160

Wylie GD, Hothem RL, Bergen DR et al (2009) Metals and trace elements in giant garter snakes (*Thamnophis gigas*) from the Sacramento Valley, California, USA. Arch Environ Contam Toxicol 56(3):577–587

Yang Z, Fang W, Lu X et al (2016) Warming increases methylmercury production in an Arctic soil. Environ Pollut 214:504–509

Zhang L, Wright LP, Blanchard P (2009) A review of current knowledge concerning dry deposition of atmospheric mercury. Atmos Environ 43:5853–5864

Current Scenario of Pb Toxicity in Plants: Unraveling Plethora of Physiological Responses

Sukhmeen Kaur Kohli, Neha Handa, Shagun Bali, Kanika Khanna, Saroj Arora, Anket Sharma, and Renu Bhardwaj

Contents

S. K. Kohli (✉) · S. Bali · K. Khanna · S. Arora · R. Bhardwaj (✉)
Department of Botanical and Environmental Sciences, Guru Nanak Dev University, Amritsar, India
e-mail: sukhmeenkohli@gmail.com; shagun88bali@gmail.com; sarojarora.gndu@gmail.com; renubhardwaj82@gmail.com

N. Handa
Department of Botany, School of Bioengineering and Biosciences, Lovely Professional University, Phagwara, Punjab, India
e-mail: neha.handa2007@gmail.com

A. Sharma
State Key Laboratory of Subtropical Silviculture, Zhejiang A&F University, Hangzhou, China
e-mail: anketsharma@gmail.com

P. de Voogt (ed.), *Reviews of Environmental Contamination and Toxicology, Volume 249*, Reviews of Environmental Contamination and Toxicology Volume 249, https://doi.org/10.1007/398_2019_25

Abbreviations

AAS	Atomic absorption spectrophotometer
ABA	Abscisic acid
ABC	ATP-binding cassettes
Ag	Silver
APOX	Ascorbate peroxidase
As	Arsenic
ATSDR	Agency for Toxic Substances and Disease Registry
Au	Gold
Ca	Calcium
CAA	Clean Air Act
CAT	Catalase
Chl a	Chlorophyll a
Chl b	Chlorophyll b
CO_2	Carbon dioxide
Cu	Copper
Cu/Zn-SOD	Copper/zinc superoxide dismutase
DHAR	Dehydroascorbate reductase
EDTA	Ethylene diamine tetraacetic acid
ETC	Electron transport chain
Fe	Iron
GB	Glycine betaine
GPOX	Glutathione peroxidase
GR	Glutathione reductase
GSH	Glutathione
GSSG	Glutathione disulfide
GST	Glutathione-S-transferase
H_2O_2	Hydrogen peroxide
Hg	Mercury
$HO^{\cdot}$	Hydroxyl radical
ICP-MS	Inductively coupled plasma-mass spectrometry
ICP-AES	Inductively coupled plasma atomic emission spectrometry
K	Potassium
MCs	Metallothionins
MDA	Malondialdehyde
MDHAR	Monodehydroascorbate reductase
Mg	Magnesium
Mn	Manganese
Mn/Zn-SOD	Manganese/zinc superoxide dismutase
Na	Sodium
Ni	Nickel
NO_3^-	Nitrate
$^1O_2$	Singlet oxygen radical

O^{2-}	Superoxide anion radical
P	Phosphorus
Pb	Lead
PCS	Phytochelatin synthetase
PCs	Phytochelatins
PGRs	Plant growth regulators
PO_4^-	Phosphate
POD	Guaiacol peroxidase
ROS	Reactive oxygen species
SDWA	Safe Drinking Water Act
SOD	Superoxide dismutase
TBARs	Thiobarbituric acid
TF	Transfer factor
TSCA	Toxic Substances Control Act
WHO	World Health Organization
Zn	Zinc

1 Introduction

Lead (Pb) (atomic number 82 and atomic weight 207.19) is bluish or silvery-gray in color with a specific gravity of 11.34 and melting point of 327.5°C (WHO 2001). It is present in solid form, usually dense, moldable, and very soft to touch with a very low electrical conductivity. It is also resistant to corrosion, making it a common metal to be used in automobiles, plastics, paints, and ceramic industries (Pinho and Ladeiro 2012; Flora et al. 2006). The common oxidative state of Pb is +2, although it has four electrons in its valence shell. Nitrate and chlorate salts of Pb have excellent solubility in water when compared to other salts (Tangahu et al. 2011). Pb is ubiquitously present in the environment, but its availability to plants is restricted due to its characteristic feature of forming complexes with soil fractions (Kos et al. 2003; Shahid et al. 2011). It is present in both marine and forest ecosystems (Sharma and Dubey 2005). It has attracted the attention of environmentalist and is widely studied in recent past as a potent environmental contaminant which is difficult to remove from contaminated habitats in the near future (Yang et al. 2000; Arshad et al. 2008; Uzu et al. 2009). Pb has been listed as a "chemical of great concern" by new European REACH regulation (EC 1907/2006; Registration, Evaluation, Authorization and Restriction of chemicals). It is also listed as the second most hazardous metal following arsenic by the Agency for Toxic Substances and Disease Registry (ATSDR 2007) based on its occurrences and toxicity in living organisms. According to ATSDR, the level of Pb in soil has enhanced by 1,000-fold over the past three centuries due to much improved analytical techniques making available more and more advanced technical tools capable to detect Pb levels that were not measurable before. The determination of Pb in soil and plants has been detected by different techniques using atomic absorption spectrophotometer (AAS), inductively coupled plasma-mass spectrometry (ICP-MS), inductively coupled plasma-atomic emission spectroscopy (ICP-AES), elemental mapping, and fluorescent sensors.

Adulteration of the soil by Pb is through breakdown of old rocks, mining and smelting activities for metal extraction, and manufacturing of paints, petroleum, and explosives (Sharma and Dubey 2005). Pb emitted from these sources settles on the soil surface. Similarly, flaking of Pb paints from the exterior of the buildings also adds substantial amount of Pb in soil (Srivastava et al. 2015). Alongside the industries, there is a substantial increase in Pb metal pollution. About 1,000 mg Pb kg^{-1} of soil has been quantified in these areas (Zaier et al. 2014). Soil and dust blown off due to wind contain naturally occurring Pb as well as Pb from industrial sources, combustion of gasoline, and deterioration of paints (Srivastava et al. 2015). Addition of Pb in air is by exhaust clouds from industries and automobiles. High amounts of tetramethyl Pb and tetraethyl Pb are added to gasoline a petroleum product to enhance its octane rating. The plants growing on highways are exposed to huge quantities of Pb as compared to plants growing away from these sites. It is still extensively used in industries and occurs abundantly in air (Punamiya et al. 2010). Similarly, Pb is added to the water bodies by battery effluents, constituents of fertilizers and pesticides, as well as sewage sludge which contains high amount of metal (Sharma and Dubey 2005). Pb enters drinking water primarily through Pb-based solders used to connect sanitary pipes in the household. Even after promotion of the restricted use of Pb by various countries, it still proves to be one of the most abundant contaminant and human health hazard (Sharma and Dubey 2005; Piotrowska et al. 2009; Grover et al. 2010).

The studies of transport of Pb into the food chain are inadequate, and furthermore assessment is needed to understand the metabolism of Pb in the food chain (Pourrut et al. 2011). The level of Pb to be considered "safe" has not yet been determined. Its threshold in soil ranges from 15 to 40 mg/kg of soil (EPA 2001). Its exposure is usually injurious to health and consequently results in irreversible effects. Lead, lead arsenate, lead sulfide, lead sulfate, lead chromate, and lead dioxide are considered as hazardous heavy metal ions and are indexed by EPA hazardous substance list of regulated toxic chemicals. Various EPA-administered laws such as Toxic Substances Control Act (TSCA), Clean Air Act (CAA), Safe Drinking Water Act (SDWA), etc. have been imposed to regulate quantity of Pb in the environment (EPA 2017). Pb has been reported to cause a wide spectrum of toxic effects on morphology, physiology, and biochemical behavior of living organisms. In plants it results in retardation of growth, root length, seed development and germination, transpiration, chlorophyll content, cell division, etc. (Gupta et al. 2009; Maestri et al. 2010). The adverse effect of Pb has been studied in various plants including *Chenopodium murale* (Sidhu et al. 2018), *Helianthus annuus* (Saleem et al. 2018), *Brassica juncea* (Kohli et al. 2018a, b), *Nicotiana tabaccum* (Maodzeka et al. 2017), *Lactuca sativa* (Silva et al. 2017), etc. The extensiveness of toxicity depends on (1) concentration of Pb, (2) duration of exposure,

(3) developmental stage of plant, and (4) organ being studied. Various mechanisms by which plants adapt to combat toxic effects of Pb include selective metal uptake, chelation with ligands, and compartmentalization (Krzesłowska et al. 2010; Maestri et al. 2010; Singh et al. 2010). For evaluation of environment quality, response of plants to Pb treatment forms an important bioindicator (Pourrut et al. 2011).

Keeping in view the long history of Pb contamination in environment and its toxic effects in plants, the present review focuses an insight into Pb toxicity in plants using physiological approaches and various strategies adopted by plant to overcome metal toxicity.

2 Lead in Soil Environment

Among heavy metal ions, Pb ions are considered to be one of the most hazardous pollutants, and their accumulation in soil and plants is of prime concern for the research due to the ill effects on human health and environment. Pb-contaminated soil has comparatively less crop yield than uncontaminated soils, consequentially creating a momentous problem for agriculture (Sharma and Dubey 2005). Pb metal is naturally present in the earth crust with its content ranging less than 50 mg kg^{-1} in soil (Pais and Jones 1997; Arias et al. 2010). The Pb content in plants varies from 2 to 200 mg kg^{-1} of dry weight, whereas in agricultural crops, its concentration is far more (Nagajyoti et al. 2010). Pb is present in the soil in four forms, i.e., (1) as free metal ions; (2) as Pb metal-inorganic compound complexes such as carbonates, chlorides, nitrates, and sulfates; (3) as adsorbed to the surface of some particles including biological matter, clay particles, and organic matter; and (4) in the form of organic ligands such as humic acids, amino acids, and fulvic acids (Tabelin and Igarashi 2009; Uzu et al. 2009; Vega et al. 2010). Sources of Pb pollution in environment have been represented in Fig. 1.

A very small fraction of Pb is available for plants to accumulate due to the fact that it forms strong complexes with colloidal particles or organic matter (Punamiya et al. 2010). Several biogeochemical features determine the behavior of Pb ion in soil in terms of solubility, adaptability, mobility, as well as bioavailability. These biogeochemical factors include pH of soil (Vega et al. 2010), concentration of Pb in soil (Lawal et al. 2010), cation-exchange capacity (Vega et al. 2010), redox properties of soil (Tabelin and Igarashi 2009), microbial activity (Arias et al. 2010), mineral availability (Dumat et al. 2006), chelating compounds (Padmavathiamma and Li 2010), and the plant species involved (Bi et al. 2010). All these factors in combination or individually contribute to the retention, bioavailability, and mobility of Pb in soil and soil solution.

Bioavailability of metal pertains to the proportion of that element in soil which is available for absorption into the living systems (humans, animals, and plants). Phytoavailability is the term used to determine portion of soil Pb available for uptake by plants (Hettiarachchi and Pierzynski 2004). The bioavailability of Pb ions is

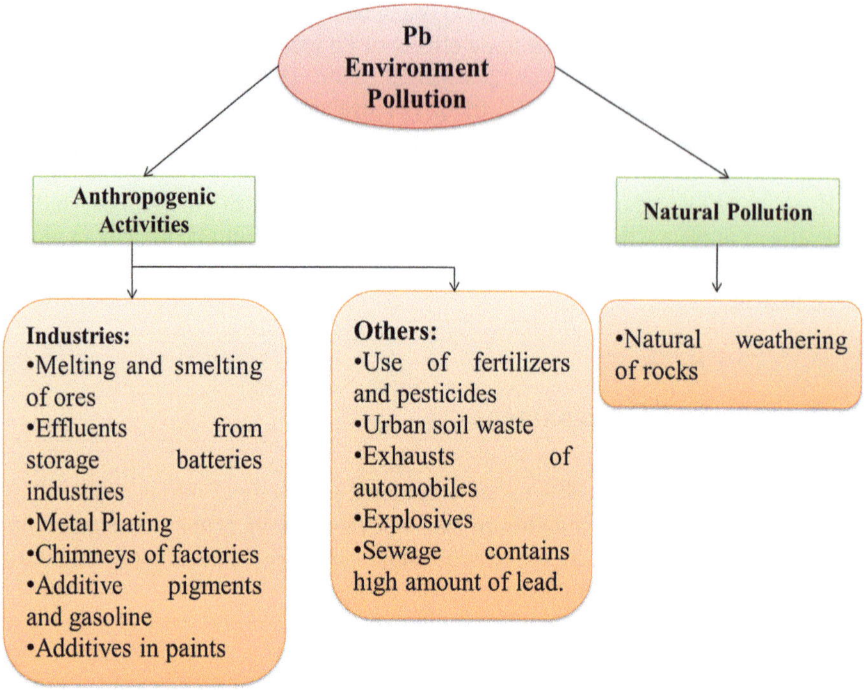

Fig. 1 Sources of Pb pollution in the environment

largely dependent upon the speciation and the concentration of free Pb ions present in the soil (Dumat et al. 2006). The reason for this lies in the fact that most of the cationic metals (specifically the free ions) are transported in plants in the form of soil solution (Punamiya et al. 2010). The amount of free Pb ions is directly dependent on various adsorption and desorption processes in which the ions are engaged (Vega et al. 2010). Pyromorphites are the most stable phosphate forms of Pb present in the soil and occur under a wide array of environmental cues (Lindsay 1979), and they occur as a common by-product of weathering of rocks in mining areas (Cotter-Howells and Thornton 1991). A report by Wu et al. (2006) suggests immobilization of water-soluble Pb due to the presence of bacterial inoculants including *Azotobacter chroococum*, *Bacillus mucilaginosus*, and *Bacillus megaterium*. This might be due to absorption of soluble Pb ions by the bacterial cell walls and sedimentation reaction with the phosphates and other anions synthesized by the bacterial immobilization. The absorption of Pb by the mineral phases (an irreversible process) represents another factor for lowered mobility and bioavailability of Pb in soil. Pb metal ions have a very high transfer rate from soil to plants, and for this reason it is broadly studied in terms of quantity, food quality, and bio-testing (Uzu et al. 2009).

3 Lead Metal Uptake and Transport in Plants

Pb metal is considered as one of the most easily sequestered metals in different plant parts (Sharma and Dubey 2005). The most significant pathway by which Pb metal is accumulated in plants is through roots from the soil/soil solution (Uzu et al. 2009). A very small fraction of Pb ions in soil solution is adsorbed on the surface of roots, and it binds to the characteristic carboxyl group of the mucilaginous uronic acid or polysaccharides of the rhizodermal cells (Seregin and Ivanov 2001). Adsorption of Pb ions on the surface of roots of higher plants is well studied in various plant species including *Brassica juncea* (Meyers et al. 2008), *Funaria hygrometrica* (Krzesłowska et al. 2009, 2010), and *Lactuca sativa* (Uzu et al. 2009).

After adsorption on the surface of rhizodermal cells, Pb metal ions are passively absorbed by the root cells and are translocated upstream along with water (Seregin et al. 2004). Uptake of lead metal ions into the plant cells is a nonselective process which is dependent upon the functioning of H^+/ATPase pump. This pump is imperative to regulate a negative potential across the membrane of rhizodermal cells (Wang et al. 2007). Pb absorption has also been reported to occur via calcium channels; hence inhibition of Pb uptake by calcium is well reported (Kim et al. 2006; Pourrut et al. 2008). It was demonstrated by Wang et al. (2007) that Ca ion channels are the main pathway for uptake of Pb. Other reports suggest uptake of Pb ions via cyclic nucleotide ion-gated channels or low-affinity cation transporters (Arazi et al. 1999; Wojas et al. 2007).

Pb is a less available and low solubility metal due to its characteristic feature to precipitate as phosphates and sulfates (Blaylock 2000). The roots of plants have an imperative ability to take up Pb as well as restrict its translocation to the aerial parts (Lane and Martin 1977). The entrance of Pb in the leaves depends upon the ability of metal to be absorbed by leaves and leaf morphology (Godzik 1996). Also, addition of certain organic chelates such as EDTA or H-EDTA at low pH increases the binding of Pb to the cell wall, in turn lowering its availability for translocation into the shoots (Jarvis and Leung 2002). The endodermis of the root cells acts as a barrier for movement of Pb from roots to the shoots. This may be a possible reason for higher accumulation of Pb in roots when compared to shoots (Verma and Dubey 2003). Moreover, Seregin et al. (2004) reported retention of significant quantity of Pb on the surface of plasmalemma rather than cell walls. They also suggested that Pb enters the injured cells along with other organic compounds including procion dyes, which do not enter the normal uninjured cells. The concentration of Pb, i.e., lethal or nonlethal, decides the pattern of accumulation of metal in the roots (Seregin et al. 2004). Few plant species tolerate Pb through forming complexes and inactivation of Pb ions (*Allium cepa*, *Zea mays*, and *Hordeum vulgare*), whereas some experience severe toxicity (*Phaseolus vulgaris* and *Brassica napus*) possibly due to hampered metabolic processes (Wierzbicka 1999).

Once Pb reaches the interior of the root cells, it may get accumulated in the root cells or translocated to the aerial parts of plants. Almost 90% of plant species accumulate Pb in root cells, while few species have been reported to sequester it in aerial parts. With exogenous supplementation of Pb, the uptake of Pb ions in roots was observed to be elevated in *Pisum sativum* plants by 96% (Małecka et al. 2008). Similarly, *Nicotiana tabbaccum* plants showed enhancement in Pb accumulation by 40-fold higher in roots when compared in aerial parts (Gichner et al. 2008). Similar increase in Pb uptake was reported by Gupta et al. (2009) in *Zea mays* plants. Pb accumulation was enhanced maximally in roots as compared to the shoots, suggesting restricted translocation of Pb ions from roots to shoots. Another observation made by Yan et al. (2010) was upregulation in Pb uptake in response to exogenous Pb treatment in root and bark. The concentration of Pb was 18.21 mg kg^{-1} and 5.39 mg kg^{-1}, respectively, in root and barks. Many possible reasons have been suggested for sequestration of Pb in roots. These include (1) intercellular precipitation of inorganic Pb salts, (2) accumulation of Pb in the plasma membrane, (3) immobilization of negatively charged pectins in cell wall, and (4) vacuolar sequestration in cortical or rhizodermal cells (Islam et al. 2007; Kopittke et al. 2007; Jiang and Liu 2010; Arias et al. 2010). Translocation of Pb ions by apoplastic movement results in the deposition of Pb in endodermis in the Casparian strips, and for further transportation, ions adopt symplastic movement (Pourrut et al. 2011). Wide array of studies suggest elevation in Pb uptake by exogenous supplementation of soil with Pb. *Triticum aestivum* plants exposed to 100 μM P treatment showed 9.6-fold enhancement in Pb uptake (Tripathi et al. 2016). Pb accumulation was recorded to be enhanced in *Echhornia crassipes* (Mart.) in response to exposure to Pb (100–1,500 mg L^{-1}) (Malar et al. 2014). Furthermore, it was revealed that levels of Pb were higher in roots than in shoots and petiole of leaves. Similar observations were made by Ashraf et al. (2017) in *Oryza sativa* plants, by Maodzeka et al. (2017) in *Nicotiana tabaccum* plants, by Leal-Alvarado et al. (2016) in *Salvinia minima*, and by Hattab et al. (2016) in *Medicago sativa* plants.

The hyperaccumulators of Pb metal include *Brassica pekinensis*, *B. juncea*, *B. pelargonium*, etc. which translocate it to the aerial parts of plant (Liu et al. 2008; Arshad et al. 2008). The translocation of Pb to the aerial parts is enhanced when it binds to EDTA (ethylene diamine tetraacetic acid), organic acids, or amino acids (Liu et al. 2008; Barrutia et al. 2010; Maestri et al. 2010) and certain microorganisms (Punamiya et al. 2010). It has been recently reported that exogenous application of chelates in the soil solution such as EDTA increases its availability and consequently enhances uptake by plants (Kos et al. 2003). A transpiration-driven uptake of Pb occurs via xylem from roots to the aerial parts of plants (Verbruggen et al. 2009; Liao et al. 2006). When Pb ions reach xylem central cylinder, it again acquires apoplastic movement and is translocated via vascular flow (Krzesłowska et al. 2010). A representation of Pb metal uptake and its transportation from soil to different parts of plant is presented in Fig. 2.

The amount of Pb absorbed into the plant cells can be measured as transfer factor (TF). This TF is determined as a ratio of Pb concentration in plant vs. Pb concentration in soil (Bi et al. 2010; Liu et al. 2010). TF is specific for specific plants and is

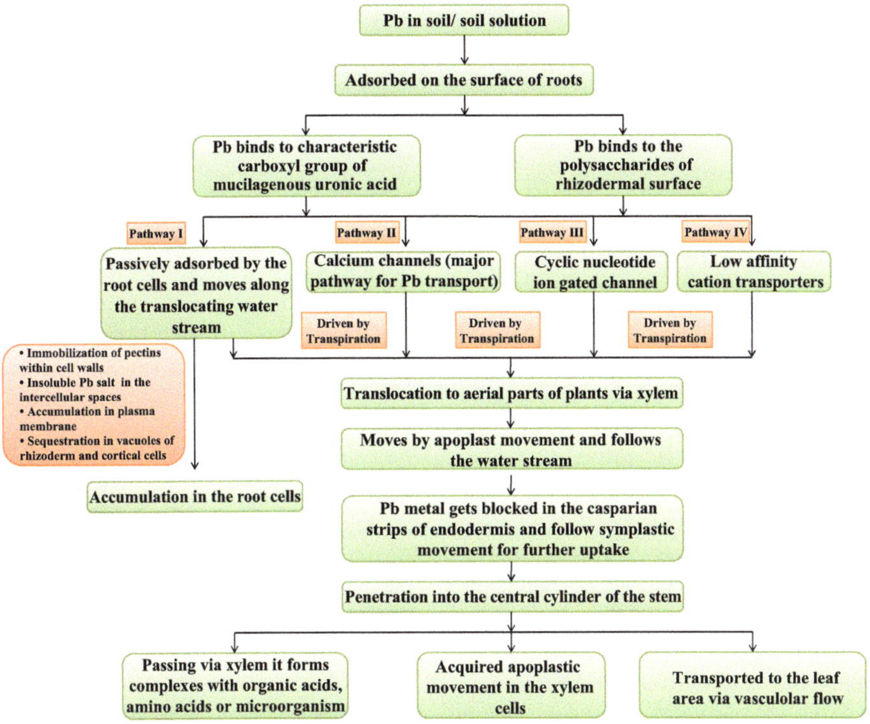

Fig. 2 Pb metal uptake and its transportation from soil to different parts of plants

dependent upon the physicochemical properties of soil (Arshad et al. 2008; Liu et al. 2010). The plant species with TF more than 1 are considered as hyperaccumulators of Pb, while plant species with TF less than 1 are termed as non-hyperaccumulators of Pb (Arshad et al. 2008). *Tradescantia spathacea* Sw. and *Chlorophytum orchidastrum* accumulate 10% and 4.48% Pb in shoots. These values are more than optimal level of >1 mg/g^{-1}, i.e., 0.1% of dry weight of plant; hence both the species are considered as hyperaccumulators of Pb (Manginsay-Enot and Sillero-Mahinay 2017). Similarly, Mohtadi et al. (2012) suggested *Noccaea caerulescens* and *Silene vulgaris* as Pb hyperaccumulators. Furthermore, a study by Alves et al. (2016) reported *Helianthus annuus*, *Fagopyrum esculentum*, and *Chrysopogon zizanioides* as potential hyperaccumulators.

4 Lead Phytotoxicity

Pb metal affects negatively some of the imperative physiological attributes of plants including germination rate of seedlings, seedling growth, dry/fresh weight of roots as well as shoots, photosynthetic efficiency, water status, mineral nutrition, and enzyme activities (Munzuroglu and Geckil 2002; Sharma and Dubey 2005). Some of the common symptoms expressed in response to Pb toxicity include necrotic lesions, chlorosis of leaves, stunted growth, and senescence (Pinho and Ladeiro 2012). Sturdy, stunted, bent, and swollen roots due to the increase in number of secondary roots are a main consequence of Pb phytotoxicity (Kopittke et al. 2007). Lowered growth in response to Pb stress is attributed to reduction in uptake and accumulation of mineral nutrients such as Fe, Mg, Cu, Zn, Mn, and P possibly due to blocked entrance of these ions and their binding to the ion carrier (Xiong 1997). Pb also significantly affects the contents of photosynthetic pigments and carbon dioxide assimilation and as a result influences the photosynthetic rate negatively (Huang and Cunningham 1996). In response to Pb toxicity, an enhancement in generation of free radicals, i.e., the reactive oxygen species (ROS) such as hydrogen peroxide (H_2O_2), superoxide anion radical (O^{2-}), hydroxyl radical ($HO^{\cdot}$), and singlet oxygen, takes place which causes oxidative stress (Jiang et al. 2010). The permeability of plasma membrane is altered in response to Pb toxicity resulting in interaction of the – SH groups, i.e., sulfhydryl groups, with the cations (Williams 1976). After Pb exposure of 48–72 h, there is disruption in functioning and physiology of plants at the cellular level including loss of cristae, swelling of mitochondria, injured plasma membrane, vacuolarized endoplasmic reticulum, and damaged nuclei (Sandalio et al. 2001). Moreover, the protein pool of the plant cell is altered in response to Pb treatment due to interaction of Pb with the proteins present in the cytoplasm (Chatterjee et al. 2004; Mishra et al. 2006). Figure 3 represents alterations in metabolic responses under the Pb stress.

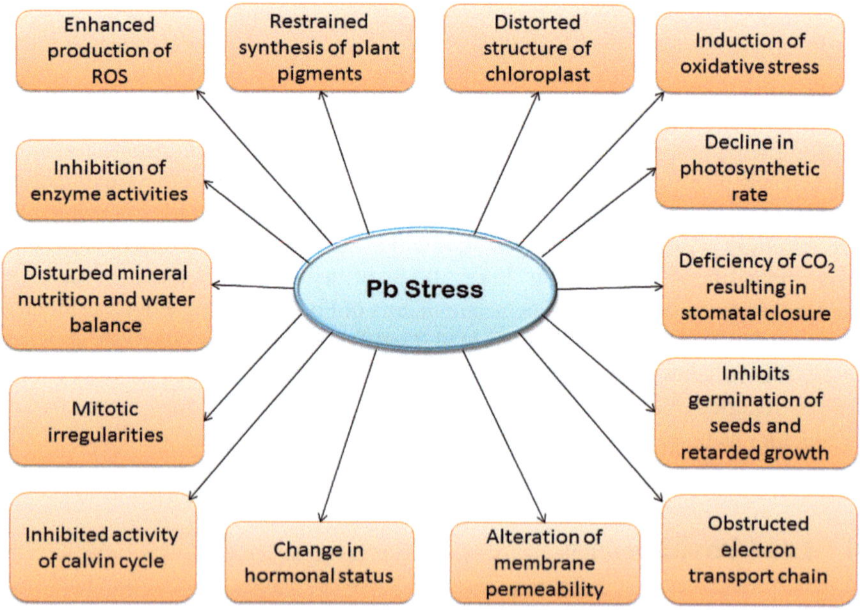

Fig. 3 Alterations in metabolic responses under the Pb stress

4.1 Growth and Water Status

Plants exposed to Pb even at micromolar concentrations show adverse effects on growth and germination (Kopittke et al. 2007). The germination of the seedlings is drastically lowered even at low concentrations of Pb (Islam et al. 2007). The germination is lowered possibly due to interference of Pb with enzymes such as amylase and proteases (Sengar et al. 2009). Growth and development of plants are significantly lowered in response to exposure to Pb (Gopal and Rizvi 2008). Moreover the growth of the aerial parts and roots is also retarded in response to low concentration of Pb treatment (Kopittke et al. 2007). Lowered plant biomass is another important symptom related to Pb exposure (Piotrowska et al. 2009; Singh et al. 2010).

Decline in growth is one of the most significant effects of Pb toxicity. The root length, shoot length, shoot fresh weight, root fresh weight, shoot dry weight, and root dry weight were reported to be lowered by 48%, 62%, 76%, 58%, 67%, and 56%, respectively, in 500 μM Pb-treated *Nicotiana tabaccum* plants in comparison to untreated plants (Maodzeka et al. 2017). Similarly, fresh weights of roots and shoots of *Medicago sativa* plants exposed to Pb treatment showed 41% and 48% decline (Xu et al. 2017). Another observation was made by Ghani et al. (2015) who reported lowered leaf length and leaf width in 20-day-old *Zea mays* plants exposed to Pb stress by 40.23% and 42.52%, respectively. Kohli et al. (2018a, b) revealed elevation in dry matter content and heavy metal tolerance index by 42.24% and 52.3%, respectively, in response to 0.75 mM Pb treatment given to *Brassica juncea* plants.

Disruption of water status in response to Pb exposure is well documented (Patra et al. 2004; Qureshi et al. 2007; Brunet et al. 2009). Reduction in growth and development of plants eventually results in retarded growth of leaf leading to reduced leaf area which results in lowered transpiration rate and elevation moisture content (Patra et al. 2004). Pb toxicity leads to altered cell turgor pressure possibly due to reduction in plasticity of cell wall. The decline in cellular metabolites like amino acids and sugars further accentuates the effect of Pb on water status (Barceló and Poschenrieder 1990). Modulation of cell turgor pressure consequently leads to change in turgidity of guard cells which interferes with opening and closing of stomata (Qureshi et al. 2007). The stomatal opening and closing are controlled by abscisic acid (ABA). The accumulation of ABA is enhanced in response to Pb toxicity in roots of stressed plants (Parys et al. 1998; Atici et al. 2005), consequently, resulting in stomatal closure (Mohan and Hosetti 1997). The enhancement in frequency of stomatal closure limits the gaseous exchange and lowers water loss by transpiration (Parys et al. 1998). Table 1 represents the effect of Pb exposure on growth and water status in different plant species.

Table 1 Effect of Pb exposure on growth and water status and photosynthetic attributes in different plant species

S. no.	Pb concentrations	Plant species	Effects on growth and water status	Reference
1.	300, 400, and 500 mg kg^{-1}	*Chenopodium murale*	Significant decline in shoot and root length, biomass of shoots and roots Total chlorophyll, chlorophyll a (Chl a), chlorophyll b (Chl b), and carotenoid levels were reduced	Sidhu et al. (2018)
2.	300, 600, and 900 mg kg^{-1}	*Helianthus annuus*	Root and shoot lengths, fresh weights, and dry weights were reduced. Carotenoid, Chl a, and Chl b contents were lowered	Saleem et al. (2018)
3.	25, 50, 75, 100 mg L^{-1}	*Ceratophyllum demersum*	Reduced level of carotenoid, Chl a, and Chl b	Dogan et al. (2018)
4.	0.25, 0.50, and 0.75 mM	*Brassica juncea*	Dry weight and heavy metal tolerance index were lowered. Water content and relative water content were enhanced	Kohli et al. (2018a)
5.	0.25, 0.50, and 0.75 mM	*Brassica juncea*	Shoot and root lengths were reduced Total chlorophyll and carotenoid levels were declined. Gas exchange attributes, viz., net photosynthetic rate, intercellular CO_2 concentration, stomatal conductance, and transpiration rate, were lowered	Kohli et al. (2017, 2018b)
6.	100, 350 and 1,500, 2,500 mg kg^{-1}	*Coronopus didymus*	Chl a, Chl b, and carotenoid contents were reduced	Sidhu et al. (2016)
7.	100, 250, and 500 μM	*Nicotiana tabaccum*	Shoot height, root length, dry weight, and fresh weight were significantly lowered Significant decline in content of SPAD value, Chl a, Chl b, and total chlorophyll	Maodzeka et al. (2017)
8.	20 mg L^{-1}	*Lactuca sativa*	Delayed germination	Silva et al. (2017)
9.	100 and 200 μM	*Hordeum vulgare*	Root length, shoot length, fresh weights, and dry weights were lowered	Arshad et al. (2017)
10.	200 mg L^{-1}	*Medicago sativa*	Shoot and root growth was inhibited	Xu et al. (2017)

(continued)

Table 1 (continued)

S. no.	Pb concentrations	Plant species	Effects on growth and water status	Reference
11.	100 μM	*Oryza sativa*	Lowered shoot, root lengths, and fresh and dry weights Lowered gas exchange parameters and pigment content	Chen et al. (2017)
12.	400, 800, and 1,200 ppm	*Oryza sativa*	Chl a and Chl b, total chlorophyll, and carotenoid contents were significantly reduced	Ashraf et al. (2017)
13.	40 μM	*Salvinia minima*	Lowering of photosynthetic rate; Chl a and Chl b and carotenoid contents were lowered	Leal-Alvarado et al. (2016)
14.	10 and 100 μM	*Medicago sativa*	Inhibition of root and shoot growth	Hattab et al. (2016)
15.	100 μM	*Triticum aestivum*	Fresh weight, dry weight, shoot length, and root length were reduced Decline in Fv/Fm ratio was observed; qP value was also lowered	Tripathi et al. (2016)
16.	25, 50, 100, and 200 μM	*Conyza canadensis*	Growth in terms of wet weights and height of plant retarded Chl a, Chl b, and total chlorophyll contents were not significantly affected	Li et al. (2016)
17.	2 mM	*Pelargonium graveolens*	Drastic retardation in growth was recorded in terms of plant height, fresh weight, dry weight, and leaf area Total chlorophyll, Chl a, Chl b, and carotenoid contents were lowered	Rao and Raghu (2016)
18.	25, 50, and 100 μM	*Brassica juncea*	Growth in terms of percentage germination, root length, shoot length, fresh weight, and dry weight was lowered Levels of total chlorophyll and carotene were reduced	Pratima and Pratima (2016)
19.	25, 50, and 100 μM	*Gossypium hirsutum*	Drastic decline in plant biomass toward leaf, stem, and root dry weight Total chlorophyll and SPAD values were significantly lowered	Bharwana et al. (2016)

(continued)

Table 1 (continued)

S. no.	Pb concentrations	Plant species	Effects on growth and water status	Reference
20.	500 μM	*Gossypium hirsutum*	Reduction in leaf morphology, viz., length, width, and petiole size	Khan et al. (2016)
21.	25, 50, 75, 100, and 150 mg kg^{-1}	*Vigna mungo*	Percentage germination, lengths of shoot and root, and the number of leaves were reduced. Similarly fresh weights, dry weights, leaf area, and number of nodules were reduced	Kumar and Jayaraman (2014)
22.	100, 200, 400, and 800 ppm	*Arachis hypogea* (cultivar K6 and K9)	Both cultivars showed reduction in growth in terms of shoot and root length. Pb induced reduction in biomass	Nareshkumar et al. (2015)
23.	200 mg kg^{-1}	*Vigna unguiculata*	Growth (number of pods per plant, number of seeds per plant and biomass) was drastically lowered	Ojwang et al. (2015)
24.	0.25 mg L^{-1}	*Pisum sativum*	Plant height, number of tendrils, and leaf length were reduced, whereas an increase in the number of leaves and leaf width was shown	Ghani et al. (2015)
25.	0.05 and 0.3 mM	*Vigna radiata*	Percentage germination and seed length were reduced	Hassan and Mansoor (2014)
26.	50 and 100 μM	*Gossypium hirsutum*	Reduction in plant height, root length, leaf area, and number of leaves per plant Decline in Chl a, Chl b, total chlorophyll, and carotenoid content and a gradual retardation in levels of net photosynthetic rate, stomatal content, and transpiration rate	Bharwana et al. (2014)
27.	100, 200, 300, 400, 500, 600, 700, 800, 900, 1,000, and 1,500 mg L^{-1}	*Eichhornia crassipes* (Mart.)	Inhibition of growth of both roots and shoots was observed. Fresh weights and dry weights were lowered, and relative water content was enhanced Total chlorophyll, Chl a, Chl b, and carotenoid contents were lowered	Malar et al. (2014)

(continued)

Table 1 (continued)

S. no.	Pb concentrations	Plant species	Effects on growth and water status	Reference
28.	1.5, 3, and 15 mM	*Triticum aestivum*	Dose-dependent reduction in growth, fresh weight, and dry weight. Total chlorophyll, Chl a, and Chl b were evidently lowered	Lamhamdi et al. (2013)
29.	1, 25, 50, 100, 200, and 500 mM	*Zea mays* L.	Percentage germination and seedling growth in terms of shoot and root length were significantly lowered. Similar reduction in fresh and dry weight of seedlings	Hussain et al. (2013)
30.	40 and 60 ppm	*Triticum aestivum*	Significant reduction in growth parameters such as shoot and root length, fresh weight and dry weight of roots and shoots, and number of tendrils Reduction in total chlorophyll, Chl a, and Chl b was observed, whereas carotenes were enhanced	Bhatti et al. (2013)
31.	50 and 100 μM	*Gossypium hirsutum*	Significant reduction in plant height, root length, leaf area, and the number of leaves was recorded. Similarly, leaf, stem, and root fresh weights and leaf, stem, and root dry weights were inhibited Significant reduction in content of Chl a, Chl b, total chlorophyll, carotenoid, and SPAD value	Bharwana et al. (2013)
32.	100, 200, 400, and 800 ppm	*Brassica juncea* L. Arawali	Seed germination and survival were reduced. The number of leaves and branches, root and shoot lengths, and fresh weights and dry weights were declined Decline in content of chlorophyll and Chl a/b ratio	Kaur et al. (2013)
33.	100, 200, and 400 μM	*Brassica napus* L.	Both varieties were observed to have lowered root length and tolerance index	Mosavian and Chaab (2012)
34.	0.1, 1, 5, 10, and 20 μM	*Sesbania exaltata*	Decline in biomass and shoot length was observed in 1 and 5 μM Pb-treated plants	McComb et al. (2012)

(continued)

Table 1 (continued)

S. no.	Pb concentrations	Plant species	Effects on growth and water status	Reference
35.	0.5 and 3.2 mM	*Lolium* sp.	Root and shoot growth was lowered by an increase in Pb concentration	Li et al. (2012)
36.	0.1, 1, 2, 3, and 4 mM kg^{-1} soil	*Jatropha curcas* L.	Leaf area, root length, and tolerance index were lowered Decline in total chlorophyll, Chl b, Chl a + b, and carotenoid content	Shu et al. (2012)
37.	50 mg L^{-1}	*Brassica juncea*	Chl a and Chl b content was lowered	John et al. (2012)
38.	2.5 mM	*Raphanus sativus*	Drastic decline in plant height, fresh weight, and dry weight Total chlorophyll content and net photosynthetic rate were significantly lowered	Anuradha et al. (2011)
39.	0.05, 0.1, 0.5, and 1 g L^{-1}	*Triticum aestivum*	Percentage germination, biomass, and percentage elongation of stem were negatively affected	Lamhamdi et al. (2010)
40.	100, 200, 400, and 800 μM	*Luffa cylindrica*	Significant decline in growth and fresh weights of plant was recorded at 800 μM concentration	Jiang et al. (2010)
41.	10, 20, and 30 ppm	*Zea mays* L.	Remarkable inhibition in root growth and shoot growth Sharp decline in level of total chlorophyll	Ghani (2010)
42.	1, 10, 50, and 100 μM	*Najus indica*	Total chlorophyll and carotenoid contents were lowered	Singh et al. (2010)
43.	10, 30, 50, and 70 μmol L^{-1}	*Albizia lebbeck* (L.) Benth	Percentage seed germination, root length, shoot length, root/shoot ratio, dry mass, and tolerance index were drastically lowered	Farooqi et al. (2009)
44.	500 μM	*Lactuca sativa* L.	Significant decline in fresh and dry weight of plants Lowering of chlorophyll, carotenoid, Chl a, and Chl b content	Đurđević et al. (2008)
45.	0.025, 0.050, 0.1, 0.15, 0.2, 0.3, 0.4, 0.5, 1, 1.5, and 2.50 mM	*Vigna unguiculata*	Decline in root and shoot growth, and fresh mass of roots and shoot was also lowered	Kopittke et al. (2007)

(continued)

Table 1 (continued)

S. no.	Pb concentrations	Plant species	Effects on growth and water status	Reference
46.	9, 10 and 11 mg L^{-1}	*Vigna mungo* PU-35 and T-9 (two inbreds)	Reduction in plant height and fresh and dry weights Reduction in chlorophyll and carotenoid content	Gupta et al. (2006)
47.	0.5, 1, 1.5, 2, 2.5, 3, 3.5, 4, 4.5, and 5 mM	*Helianthus annus*	Decline in shoot and root lengths was observed. Similarly, shoot length and dry matter stress tolerance index were also lowered	Azhar et al. (2006)
48.	100 mg L^{-1}	*Sinapis alba*	Inhibitory effect on root elongation. Lowered contents of total chlorophyll, Chl a, Chl b, Chl a/b ratio, Chl (a + b)/carotenoid ratio, and carotenoid	Fargasova (2004)
49.	500 and 1,000 μM	*Oryza sativa*	Decline in lengths of shoot and roots and fresh weight of seedlings	Verma and Dubey (2003)

4.2 Photosynthetic Efficiency

Photosynthesis involves conversion of light energy to chemical energy which is imperative for a huge amount of metabolic processes (Ashraf and Harris 2013; Taiz and Zeiger 2010; Pan et al. 2012). Disrupted photosynthetic efficiency and modified ultrastructure of chloroplast are primary symptoms of metal toxicity (Garg and Aggarwal 2011). Heavy metals including Pb, Mn, Cu, and Fe affect deleteriously both light and dark reactions of photosynthesis either directly or indirectly. The direct effects include alterations in light reactions through changes in O_2 evolution, phosphorylation, and NADP reduction, whereas the indirect effects include pigment synthesis and degradation and competition between toxic metal ions and essential metal ions (Aggarwal et al. 2012). One of the most apparent effects of HM toxicity on the photosystems (PS I and PSII) is inhibition of O_2 evolution. Furthermore, it was suggested that these metal ions inhibit the donor as well as acceptor sites of PS II. On the acceptor sites of PS II, quinine acceptor Q_A, non-heme iron region, pheophytin-Q_A-iron region, and secondary quinine acceptors are identified as a target of metal-induced inhibition (Aggarwal et al. 2012).

The possible reasons for lowered photosynthetic efficiency in response to Pb toxicity might be due to (1) strong affinity of Pb ions to bind to N and S ligands resulting in damage of the structure of chloroplast (Islam et al. 2007), (2) disrupted electron transport chain (ETC) (Qufei and Fashui 2009), (3) degradation of chlorophyll by enhancing the activity of chlorophyllase (Liu et al. 2008), (4) inhibition and substitution of metal nutrients such as Mg and Fe by Pb ions, (5) lowered CO_2 levels due to closing of stomata (Romanowska et al. 2006), (6) lowered activity of enzymes such as ferredoxin-$NADP^+$ reductase and delta-aminolevulinic acid dehydratase (Gupta et al. 2009), (7) inhibition of catalytic enzymes of Calvin cycle (Liu et al. 2008), and (8) decline in content of carotenoid and plastoquinone (Jing et al. 2007; Liu et al. 2008). The process of chlorophyll degradation is triggered by enzymes such as chlorophyllase, red chlorophyll catabolite reductase, and Mg-dechelatase. These enzymes catalyze the disassociation of chlorophyll into Mg, phytol, and porphyrin ring. These enzymes are also responsible for enhanced chlorophyll bleaching, i.e., loss of green color (Harpaz-saad et al. 2007). The content of Chl b is more affected by Pb treatment when compared to Chl a (Xiong and Wang 2005). Inhibition of photosynthetic activity by Pb is well documented (Liu et al. 2008; Piotrowska et al. 2009; Singh et al. 2010). The reduction in photosynthetic activity is very often considered as a more sensitive indicator of toxicity to decline in pigment content (Harpaz-saad et al. 2007). Pb-exposed plants show a reduction in the quantity of grana stacks together due to lowered amount of stroma as well as decline in starch grains levels. Pb toxicity also resulted in altered composition of lipids of thylakoid membrane (Stefanov et al. 1995). The effects of Pb exposure on photosynthetic attributes in different plant species are tabulated in Table 1.

4.3 *Oxidative Damage*

One of the most significant phytotoxic effects of Pb is considered to be the generation of oxidative stress in response to production of ROS. The redox status of cell is altered in response to ROS production. Several different forms of ROS are produced including superoxide anion ($O_2^{\cdot -}$), hydrogen peroxide (H_2O_2), hydroxyl ions ($^{\cdot}OH$), and singlet oxygen species ($^1O_2$). These ROS pose a severe challenge to maintain cellular metabolism when produced in large quantities (Liu et al. 2008; Yadav et al. 2010). ROS are molecules with one pair of unpaired electrons making them highly reactive entities, which react with several cellular components in a non-specific manner (Ashraf 2009). Free radicals are also produced in normal cells during metabolic processes of chloroplast. The generation of ROS in plants under the effect of Pb is well documented (Grover et al. 2010; Singh et al. 2010; Yadav et al. 2010). Production of ROS is directly dependent upon the following characteristics: (1) type of metal, (2) form in which metal exists, (3) plant species affected, and (4) duration of exposure (Wang et al. 2007). They rapidly oxidize several important biomolecules including lipids, proteins, and nucleic acid. The effect of this attack is irreparable and causes metabolic dysfunctions and eventually cell death (Yadav et al. 2010). Exogenous supplementation of plants with Pb results in drastic elevation in MDA content. The MDA content was enhanced in *Lycopersicon esculentum* plants (Kumar et al. 2017). MDA levels were enhanced by 15% and 37% in leaf and root tissues of *Eicchornia crassipes* (Mart.) plants exposed to 400 mg L^{-1} Pb treatment (Malar et al. 2014). Similarly, elevation in lipid peroxidation was reported in

Gossypium hirsutum plants under Pb concentration (50 μM and 100 μM Pb) (Bharwana et al. 2013). MDA levels were also reported to be enhanced from 16.07 μmol mg^{-1} protein to 28.23 μmol mg^{-1} protein in 500 μM Pb-treated *Gossypium hirsutum* plants.

Pb toxicity has been observed to induce lipid peroxidation in many plant species resulting in reduced levels of saturated fatty acids and enhanced contents of unsaturated fatty acids in the cell membrane (Singh et al. 2010). A marked alteration is caused by Pb in lipid composition of different cellular membranes (Piotrowska et al. 2009; Grover et al. 2010; Singh et al. 2010). Very high susceptibility to ROS is observed in poly-saturated fatty acids and their esters present in lipids (Dey et al. 2007; Gupta et al. 2009). ROS remove hydrogen from the fatty acids and result in the formation of lipid radicals and aldehydes, consequently leading to distortion of lipid bilayer of membranes (Mishra et al. 2006). The polyunsaturated fatty acids have bis-allylic hydrogen present on their surface. These bis-allylic hydrogen undergo oxidation in the presence of ROS. Oxidation involves the following steps: (1) formation of lipid radicals, i.e., the initiation step, and (2) biomolecular reaction of lipid peroxyl radicals which results in the formation of non-radicals, i.e., progression and termination step (Gurer and Ercal 2000; Bhattacharjee 2005). These alterations in the lipid membrane cause abnormal changes in the ultrastructure of the cell membrane (Gupta et al. 2009) and other organelles such as peroxisomes and mitochondria (Liu et al. 2008). Table 2 shows the effect of Pb treatment on ROS and MDA levels in different plant species.

4.4 Proteins

High concentrations of Pb have been reported to lower the protein pool of the cytoplasm (Mishra et al. 2006; Piotrowska et al. 2009). Protein content of Pb-stressed cell is lowered possibly due to enhancement in oxidative stress and production of ROS (Gupta et al. 2009), enhancement in activity of certain enzymes including ribonucleases (Gopal and Rizvi 2008), overutilization of proteins as a Pb detoxification mechanism (Gupta et al. 2009), and modulation of contents of free amino acids (Gupta et al. 2009). There are reports of few amino acids such as proline, to be enhanced under Pb stress (Qureshi et al. 2007). Low or nonlethal concentration of Pb results in elevation of protein content (Mishra et al. 2006), which might be a defense mechanism of plants against Pb stress (Gupta et al. 2010). Protein levels were lowered by 9.13% in shoots and by 20.34% in roots when exposed to 1 mM Pb concentration in *Zea mays* plants (Hussain et al. 2013). The total soluble protein levels in *Jatropha curcas* plants showed drastic reduction. They further suggested that protein content was initially enhanced at 1 mM concentration and was lowered in 4 mM Pb-treated plants (Shu et al. 2012). The effect of Pb exposure on protein content in different plant species is given in Table 2.

Table 2 Effect of Pb exposure on ROS and MDA and protein levels in different plant species

S. no.	Pb concentration	Plant species	Effects on content of ROS and MDA	Reference
1.	300, 400, 500 mg kg^{-1}	*Chenopodium murale*	Contents of superoxide anion, H_2O_2, and MDA were enhanced	Sidhu et al. (2018)
2.	300, 600, and 900 mg kg^{-1}	*Helianthus annus*	MDA levels were elevated	Saleem et al. (2018)
3.	25, 50, 75, 100 mg L^{-1}	*Ceratophyllum demersum*	Levels of lipid peroxidation were enhanced Dose-dependent decline in protein level	Dogan et al. (2018)
4.	0.25, 0.50, 0.75 mM	*Brassica juncea*	H_2O_2, superoxide anion levels, and MDA content were enhanced	Kohli et al. (2017)
5.	100 μM	*Solanum lycopersicum*	MDA levels were enhanced Protein content was minimally enhanced	Kumar et al. (2017)
6.	400, 800, and 1,200 ppm	*Oryza sativa*	H_2O_2 levels and MDA content were enhanced	Ashraf et al. (2017)
7.	2 and 50 mg kg^{-1}	*Salix matsudana*	Total soluble protein levels were lowered	Tang et al. (2017)
8.	10, 250, and 500 μM	*Nicotiana tabacum*	MDA content was observed to elevate in all genotypes in a dose-dependent manner	Maodzeka et al. (2017)
9.	1, 2, 4, 6, and 8 mg kg^{-1}	*Lemna minor*	Reduced protein levels	Jayasri and Suthindhiran (2017)
10.	500 μM	*Gossypium hirsutum*	Soluble protein levels were elevated	Khan et al. (2016)
11.	40 μM	*Salvinia minima*	Enhancement in membrane damage	Leal-Alvarado et al. (2016)
12.	10 and 100 μM	*Medicago sativa*	TBARs content in roots and shoots enhanced pronouncedly	Hattab et al. (2016)
13.	100 μM	*Triticum aestivum*	Shoot and root contents of MDA and H_2O_2 were elevated	Tripathi et al. (2016)
14.	25, 10, and 100 μM	*Gossypium hirsutum*	Elevation in H_2O_2 and MDA content Protein level was significantly lowered	Bharwana et al. (2016)

(continued)

Table 2 (continued)

S. no.	Pb concentration	Plant species	Effects on content of ROS and MDA	Reference
15.	100, 200, 400, and 800 ppm	*Arachis hypogea* L. (cultivar K6 and K9)	Enhanced localization of $O_2^{\cdot -}$ and H_2O_2 was observed by biochemical as well as histological analysis. MDA content in roots as well as leaves of both cultivars was elevated	Nareshkumar et al. (2015)
16.	5 μM	*Vicia faba*	Levels of TBARs and H_2O_2 content were enhanced	Shahid et al. (2015)
17.	50 and 100 μM	*Gossypium* plants	Levels of electrolyte leakage and H_2O_2 content were elevated Levels of soluble protein in roots and leaves were significantly lowered	Bharwana et al. (2014)
18.	0.05 and 3 mM	*Vigna radiata*	No significant influence on MDA content 0.3 mM Pb resulted in lowered levels of soluble protein	Hassan and Mansoor (2014)
19.	100, 200, 300, 400, 500, 600, 700, 800, 900, 1,000, and 1,500 mg L^{-1}	*Eichhornia crassipes* (Mart.)	MDA content was elevated	Malar et al. (2014)
20.	50 and 100 μM	*Gossypium hirsutum*	Electrolyte leakage, MDA, and H_2O_2 levels were recorded to be elevated Content of soluble protein was significantly lowered	Bharwana et al. (2013)
21.	100, 200, 400, and 800 ppm	*Brassica juncea arawali*	Soluble protein content was lowered	Kaur et al. (2013)
22.	1.5, 3, and 15 mM	*Spinacia oleracea* and *Triticum aestivum*	Elevation in protein content in both plants	Lamhamdi et al. (2013)
23.	1, 25, 50, 100, 200, and 500 mM	*Zea mays*	Reduction in protein content	Hussain et al. (2013)
24.	100, 200, and 400 μM L^{-1}	*Brassica napus* (two variety: Hayola308 and RG5003)	Remarkable elevation in MDA was recorded in aerial parts of Hayola variety	Mosavian and Chaab (2012)
25.	0.50 and 3.2 mM	*Lolium* sp.	Substantial increase in MDA content. Soluble protein content was observed to enhance	Li et al. (2012)

(continued)

Table 2 (continued)

S. no.	Pb concentration	Plant species	Effects on content of ROS and MDA	Reference
26.	2.5 mM	*Raphanus sativus*	Drastic increase in MDA content	Anuradha et al. (2011)
27.	0.1, 2, 3, and 4 mM kg^{-1}	*Jatropha curcas*	Protein level was lowered	Shu et al. (2012)
28.	0.05, 0.1, 0.5, and 1 g L^{-1}	*Triticum aestivum*	MDA content drastically elevated Soluble protein content was significantly elevated	Lamhamdi et al. (2010)
29.	1, 10, 50, and 100 μM	*Najus indica*	Elevation in MDA and H_2O_2 levels Enhancement in protein level	Singh et al. (2010)
30.	0.5 mM	*Raphanus sativus*	Higher levels of TBARs in leaves and hypocotyls	Teklić et al. (2008)
31.	500 and 1,000 μM	*Oryza sativa*	Increase in levels of TBARs and superoxide anion content	Verma and Dubey (2003)

4.5 Mineral Nutrition

Imbalance of mineral nutrition is caused in growing plants as a result of high levels of Pb in the soil vicinity. Kabata-Pendias (1993) reported a significant alteration in the contents of nutrients as well as internal ratio of nutrients in plants under Pb stress. In most of the cases, it was reported that Pb blocked the entry of cations including Ca, Mg, K^+, Cu, and Fe^{3+} and certain anions like NO_3^- and PO_4^- in the roots. Pb blocks the entry of these ions from the absorption sites on the root surface (Godbold and Kettner 1991). The observation of several studies suggests that uptake of nutrients is drastically affected by the presence of Pb (Sharma and Dubey 2005; Gopal and Rizvi 2008). Several reports suggest replacement or decline in levels of divalent ions such as Ca^{2+}, Mg^{2+}, Mn^{2+}, Zn^{2+}, and Fe^{2+} by Pb exposure in *Brassica oleracea* (Sinha et al. 2006), in leaves of *Zea mays* (Seregin et al. 2004), *Vigna unguiculata* (Kopittke et al. 2007), and *Raphanus sativus* (Gopal and Rizvi 2008). The reduced nutrient uptake in the presence of Pb ions might have resulted from competition of Pb ions with other divalent ions of similar atomic size or altered physiological activity (Sharma and Dubey 2005). They also suggested that there occurs a strong interaction of K^+ ions with Pb due to similarity of the radii, i.e., K^+, 1.33 Å, and Pb^{2+}, 1.29 Å. As a result, these two ions compete for entry into the plant cells via similar potassium channels. In a similar manner, Pb also affects K^+ ATPase and –SH groups of the cell membrane proteins resulting in efflux of K^+ from the roots. There is a reduction observed in the levels of inorganic nitrogen due to lowered activity of nitrate reductase enzyme (Xiong and Wang 2005; Sengar et al. 2009). Ca, Mg, Zn, P, and Ni contents were lowered in response to 100 μM Pb treatments in *Triticum aestivum* (Bharwana et al. 2016). Similarly, Ca, Mg, P, K, and Na levels were reduced in *Spinacia oleracea* and *Triticum aestivum* in response to Pb (1.5, 3, and 15 mM) treatment (Lamhamdi et al. 2013). K^+ and Na^+ ion content declined in roots as well as shoots when exposed to Pb (40 and 60 ppm) in *Triticum aestivum* plants (Bhatti et al. 2013). Another report suggested decline in shoot tissue concentration of Mg, K, and Cu, and slight elevation in the contents of P and S was recorded in *Vigna unguiculata* plants exposed to Pb (0.025, 0.050, 0.1, 0.15, 0.2, 0.3, 0.4, 0.5, 1, 1.5, and 2.5 μM) treatments (Kopittke et al. 2007). Moreover, Gupta et al. (2006) observed progressive reduction in N content in two inbreds (PU-35, T-9) of *Vigna mungo* in response to Pb treatment.

4.6 Phenolic Compounds

The plants produce a wide range of secondary metabolites, and one of the most imperative groups of these metabolites includes phenolic compounds. Phenolic compounds are characterized by the presence of at least one aromatic ring (C6) bearing one or more hydroxyl groups. The primary precursor of synthesis of phenolic compounds is L-phenylalanine ammonia lyase (PAL) (Dixon and Paiva 1995). Phenolic compounds have broad-spectrum role in modulation of physiological responses in plants. Simple phenols including coumaric acid and cinnamic acid act as a precursor for synthesis of complex phenolic compounds like flavonoids, lignins, tannins, and anthocyanins (Šebestík et al. 2011). Elevation in metabolism of phenylpropanoid and levels of phenolic compounds is observed in plants under the influence of varied environmental factors and stresses (Sakihama and Yamasaki 2002). These phenolic compounds have an activity similar to that of antioxidants under stressful conditions, possibly due to their electron-donating nature (Michalak 2006). The antioxidative action of phenolic compounds is also attributed to their tendency to bind or chelate with the metal ions, as phenolic compounds possess a hydroxyl and carboxyl group which binds to metal ion (Decker 1997). Similarly, polyphenols have also been reported to possess an ideal structure for scavenging of ROS under in vitro conditions in rice plants (Rice-Evans et al. 1997). Polyphenols also display a wide variety of responses in plants including resistance to microbial infection, metal stress, and solar radiation. It has been suggested by several workers that phenol metabolism is stimulated in response to metal stress (Sharma et al. 2011; Devi et al. 2014).

Phenolic compounds are also characterized as a significant biomarker of metal stress. Enhanced level of phenolic compounds has been reported in several plant species including wheat under Ni stress (Dai and Mumper 2010), maize under Al stress (Winkel-Shirley 2002), common bean under Cd stress, and *Phyllanthus tenellus* under Cu stress (Dai and Mumper 2010). Also, Lavid et al. (2001) suggested direct chelation of Cr, Hg, and Pb ions with phenolic compounds in *Nymphaea alba* plants. Endogenous levels of total phenols and flavonoids were reported to be elevated in *Tithonia diversifolia* plants exposed to Pb acetate suggesting imperative role in adaptation to metal toxicity (Olivares 2003). Another observation made by Muszyńska et al. (2014) showed that *Hippophae rhamnoides* L. grown in waste heaps obtained after Pb ore flotation in Bukowno near Olkusz (Southern Poland) had elevated levels of phenols, phenylpropanoids, flavanols, and anthocyanin revealing their antioxidant activity. Moreover, Wiszniewska et al. (2017) suggested that exogenous application of organic amendments specifically phenolic acids resulted in improvement in tolerance of *Daphne jasminea* plants to Pb stress by enhancing the radical scavenging activity. Another report demonstrates involvement of polyphenols in ROS generation under metal stress at cellular level. Tripathi et al. (2016) reported reduction in levels of total phenolic content in *Triticum aestivum* plants exposed to Pb (100 μM) metal stress. Initially the content of total polyphenols was reported to enhance, followed by reduction at higher concentration of Pb (10, 20, 30 and 40 mM) in *Brassica juncea* plants (Pratima and Pratima 2016).

5 Pb Tolerance in Plants

Plants respond to toxic effects of Pb in various ways including judicious uptake of metal, metal chelation to the root surface and cell wall, and activation of antioxidative defense cascade (Pourrut et al. 2011). Plants have numerous strategies to defend the toxic effects of heavy metals. The primary defense mechanism is termed as passive mechanism which involves physical avoidance of metal entry into the plant system. The secondary mechanism is induction of various metal transporters, antioxidants, and metal-chelating compounds. The tertiary defense mechanism involved activation of antioxidative enzymes (Rossato et al. 2012; Pourrut et al. 2011).

5.1 Passive Mechanism

Entry of even a very low concentration of Pb from the root cell surface results in its interaction with the cellular components and eventually leads to increased thickness of cell wall as defense response (Krzesłowska et al. 2010; Krzesłowska 2011). Pectin is an imperative cell wall component. Pb metal ions form complexes with pectin's carboxylic group. This interaction or formation of pectin-Pb complex formation is observed as a significant method by which plant counters Pb toxicity (Meyers et al. 2008; Jiang and Liu 2010). There are several reports which suggest binding of Pb with JIM5-P which are monoclonal antibodies (low-methyl esterified pectins, i.e., pectin epitope JIM5-P) which act as a physical barrier that restricts the uptake of Pb ions into the plant cells such as *Funaria hygrometrica* (Krzesłowska et al. 2010). They further suggested that Pb binds to JIM5-P within the cellular compartments and not on the surface and is taken up or re-mobilized by the process of endocytosis.

Moreover, they also demonstrated formation of similar wall thickening in the diverse groups of plant species such as *Arabidopsis* hybrids, *Populus tremuloides*, and *Lemna trisulca*. They further state that these thickenings are formed on the tip of growing cells such as root hairs, root caps, and other meristematic regions. All the cell thickenings under observation were found to be rich in JIM5-P and were found to sequester large quantities of Pb ions. Hence, it was concluded that cell wall sequestration and compartmentalization of Pb metal prevent protoplast toxicity.

5.2 Activation of Antioxidative Defense Response

Plants have developed numerous mechanisms to combat toxicity of heavy metals. One of the primary defense mechanisms is avoidance of metal entry into the cell and the secondary defense mechanism involves the activation of antioxidative defense system which includes modulation of activities of antioxidative enzymes and contents of antioxidants to counter high levels of ROS caused by heavy metals (Rossato et al. 2012). The antioxidative defense system is comprised of SOD (superoxide dismutase), CAT (catalase), APOX (ascorbate peroxidase), glutathione (GSH), vitamin C (ascorbic acid), vitamin E (tocopherol), and carotenoids. These components are evenly distributed throughout the plant cells (Pinho and Ladeiro 2012). The distribution of antioxidants as suggested by Scandalios (2005) in plants cells is (1) 73% in vacuoles (peroxidases, ascorbates, and glutathione), (2) 17% in chloroplast (Cu/Zn-SOD, MDHAR, DHAR, GR, GSH, vitamin E, vitamin C, and APOX), (3) 5% in cytosol (APOX, Cu/Zn-SOD, CAT, POD, GSH, vitamin C, GR, and Mn/Zn-SOD), (4) 4% in apoplast (POD and vitamin C) and 1% in mitochondria (CAT, GSH, GR, Mn-SOD, and MDHAR) and peroxisome (CAT and Cu/Zn-SOD).

Exposure of Pb results in elevation in activities of several enzymes including GPOX, SOD, APOX, CAT, GST, and GR (Singh et al. 2010; Gupta et al. 2009, 2010; Brunet et al. 2009). The enhancement in activities of enzymes is probably due to modulation of gene expression or blocking of certain enzyme inhibitors (Seregin and Ivanov 2001). In response to heavy metal toxicity, ROS are produced which are scavenged by various antioxidative enzymes. The primary defense against ROS is countered by metalloenzymes present in the cellular compartment, i.e., SOD (Mishra et al. 2006). SOD stimulates dismutation of superoxide radicals to H_2O_2 and molecular oxygen, thus maintaining the levels of superoxide anion in the cytosol (Islam et al. 2008). Moreover, H_2O_2 is an important oxidant and requires to be removed quickly from the plant system as it is highly toxic. H_2O_2 is scavenged by various enzymes including APOX, GR, and CAT (Mishra et al. 2006). Halliwell-Asada antioxidative defense pathway demonstrates significant participation of GSH and GR in scavenging of H_2O_2 in the plant cells (Piechalak et al. 2002). Activation of antioxidative enzymes occurs by increased Pb toxicity or indirectly by increasing the concentration of substrates of antioxidative enzymes (Islam et al. 2008).

Similar to other metals, Pb also affects the activity of a wide range of enzymes of varied metabolic pathways. At a concentration of 2×10^{-5}–2×10^{-4} M, Pb has been reported to inhibit several enzymes, and this concentration is termed as inactivation constant (Ki) (Seregin and Ivanov 2001). Inhibition of enzymes by Pb metal is dependent upon several factors including plant species, metal type, duration, and level of treatment (Islam et al. 2008; Singh et al. 2010). The inactivation of enzymes occurs in response to interaction between protein catalytic site and results in disruption of its structure. Pb metal causes similar effects by binding to protein-COOH groups (Gupta et al. 2010). It was suggested by Seregin and Ivanov (2001) that Pb exposure enhances the activity of several enzymes by modulating the gene expression and also by the activity of enzyme inhibitors. As a result of enhanced levels of antioxidative enzymes in response to metal toxicity, the levels of ROS are lowered (Mishra et al. 2006). Few reports of alterations in antioxidative defense system in response to Pb treatment have been tabulated in Table 3.

5.3 Heavy Metal Transporter Proteins

Heavy metal transporter proteins play a pivotal part in metal detoxification in plant cells via excursion of metal ions into the extracellular species (Meyers et al. 2008; Maestri et al. 2010). The heavy metal transporter proteins in plants are classified into metal-efflux proteins and metal-uptake proteins. The metal-uptake transporters bind to essential heavy metal and carry them into the cytoplasm. Metal-efflux transporters are categorized as metal detoxification proteins. These proteins efflux excess and toxic heavy metal from the cytoplasm also help in compartmentalization of metals (Qian et al. 2016). A human DMT1 (divalent metal transporter 1) has been reported to be expressed in yeast cells and is also absorbed to transport Pb ions by pH-dependent phenomena in plants (Bressler et al. 2004). Moreover several ATP-binding cassette (ABC) carriers including AtADPR12 and AtATM3 in *Arabidopsis* were found to be involved in resistance or enhanced tolerance to Pb (Kim et al. 2006). Transcriptomic analyses of several genes of these carriers suggest upregulation of these carriers in response to Pb toxicity (Liu et al. 2008).

Sequestrations of metals in cellular compartment are considered as a significant aspect of plant metal detoxification and plant nutrient homeostasis (Maestri et al. 2010). Pb is also reported to bind to several organic molecules and is compartmentalized in various organelles including dictyosome vesicles (Malone et al. 1974), plasma tubules (Wierzbicka 1999), endoplasmic reticulum vesicles (Wierzbicka et al. 2007), and vacuoles (Małecka et al. 2008; Meyers et al. 2008). PCs and MTs are most imperative metal-binding ligands in the plant cells. These chelating compounds belong to various classes of cysteine-rich metal-ligating protein biomolecules. PCs are most often reported as metal toxicity-protecting protein in plants; as stated earlier they are low-molecular-weight proteinaceous metal-ligating compounds (Maestri et al. 2010). They are found to form mercaptide bonds with the metal ions (Gupta et al. 2010; Jiang and Liu 2010; Yadav et al. 2010). These are biologically active compounds which prevent plant cells from oxidative damage (Gupta et al. 2010).

Table 3 Effect of Pb exposure on antioxidative defense system in different plant species

S. no.	Pb concentration	Plant species	Effects on antioxidative defense system	Reference
1.	300, 400, and 500 mg kg^{-1}	*Chenopodium murale*	Enhancement in SOD, POD, and CAT enzyme activities	Sidhu et al. (2018)
2.	300, 600, and 900 mg kg^{-1}	*Helianthus annus*	Elevation in SOD, CAT, APOX, and GR enzyme activities	Saleem et al. (2018)
3.	0.25, 0.50, and 0.75 mM	*Brassica juncea*	SOD, POD, and CAT activities were enhanced, whereas GR and GST activities were lowered	Kohli et al. (2018b)
4.	100 μM	*Oryza sativa*	The activities of SOD, APOX, and GR were increased. Antioxidants including ascorbic acid and GSH levels were reduced	Chen et al. (2017)
5.	25 and 50 mg kg^{-1}	*Salix mutsudana*	Enzyme activities of APOX and CAT were enhanced, while SOD, GPOX, and GR activities were lowered	Tang et al. (2017)
6.	0.25, 0.50, and 0.75 mM	*Brassica juncea*	SOD, POD, and CAT activities were enhanced, whereas GR and GST activities were lowered	Kohli et al. (2017)
7.	400, 800, and 1,200 ppm	*Oryza sativa*	SOD, POD, CAT, and APOX activities were recorded to be enhanced. GSH and GSSG were also elevated	Ashraf et al. (2017)
8.	10, 250, and 500 μM	*Nicotiana tabacum*	CAT and POD enzyme activities were increased	Maodzeka et al. (2017)
9.	10 and 100 μM	*Medicago sativa*	GR and APOX activities were enhanced. Also SOD and its four isomers' (Cu-SOD, Zn-SOD, Mn-SOD, Fe-SOD) activities were elevated	Hattab et al. (2016)
10.	25, 10, and 100 μM	*Gossypium hirsutum*	Activities of SOD, POD, and APOX were elevated in leaves as well as roots	Bharwana et al. (2016)
11.	25, 50, 100, and 200 μM	*Conyza canadensis*	SOD, CAT, and POD activities were recorded to be higher in leaves. SOD and CAT activities were lowered in roots as well but no significant change in POD activity	Li et al. (2016)
12.	10, 20, 30, and 40 ppm	*Brassica juncea*	Ascorbate and glutathione were lowered. SOD, CAT, and POD were also reduced	Pratima and Pratima (2016)

(continued)

Table 3 (continued)

S. no.	Pb concentration	Plant species	Effects on antioxidative defense system	Reference
13.	100, 200, 400, and 800 ppm	*Arachis hypogea* (cultivar K6 and K9)	Ascorbic acid content was significantly elevated. SOD, APOX, GPOX, GR, and GST were enhanced	Nareshkumar et al. (2015)
14.	50 and 100 μM	*Gossypium* plants	Higher activity of SOD, POD, CAT, and APOX was recorded in both roots and leaves	Bharwana et al. (2014)
15.	0.05 and 3 mM	*Vigna radiata*	Activities of GPOX and CAT enzymes were enhanced	Hassan and Mansoor (2014)
16.	100, 200, 300, 400, 500, 600, 700, 800, 900, 1,000, and 1,500 mg L^{-1}	*Eichhornia crassipes* (Mart.)	SOD, APOX, CAT, and peroxidase enzyme activities were enhanced	Malar et al. (2014)
17.	50 and 100 μM	*Gossypium hirsutum*	Activities of SOD, POD, and APOX were recorded to be enhanced	Bharwana et al. (2013)
18.	0.50 and 3.2 mM	*Lolium* sp.	CAT, APOX, and POD activities were enhanced, whereas SOD activity was lowered. Similar results were recorded in gene expression studies of Mn-SOD, Cu/Zn-SOD, and Fe-SOD	Li et al. (2012)
19.	50 mg kg^{-1}	*Brassica juncea*	Increase in ascorbic acid content was observed	John et al. (2012)
20.	2.5 mM	*Raphanus sativus*	Activities of peroxidase, APOX, and SOD enzymes were enhanced. CAT and GR activities were lowered	Anuradha et al. (2011)
21.	0.05, 0.1, and 0.5 and 1 g L^{-1}	*Triticum aestivum*	SOD, CAT, APOX, POD, and GST enzyme activities were enhanced	Lamhamdi et al. (2010)
22.	0.5, 1, 2, 3, and 4 mM kg^{-1} soil	*Jatropha curcas*	SOD and POD activities were recorded to be lowered in response to Pb stress. CAT activity was also elevated	Shu et al. (2012)
23.	100, 200, 400, and 800 μM	*Luffa cylindrica*	Significant elevation in enzyme activity of SOD, POD, and PAL	Jiang et al. (2010)
24.	1, 10, 50, and 100 μM	*Najus indica*	Reduced GSH and oxidized GSSG was elevated	Singh et al. (2010)
25.	0.5 mM	*Raphanus sativus*	Total peroxidase activity is higher in hypocotyls and leaves. CAT activity to be more in leaves than hypocotyls	Teklić et al. (2008)

(continued)

Table 3 (continued)

S. no.	Pb concentration	Plant species	Effects on antioxidative defense system	Reference
26.	500 and 1,000 μM	*Oryza sativa*	SOD, CAT, GPOX, and APOX were enhanced. GR activity elevated	Verma and Dubey (2003)

Cysteine and glutathione are the nonenzymatic components of plant defense system. An increase in levels of cysteine has been reported in *Arabidopsis* plants under Pb stress (Liu et al. 2008). On the other hand, glutathione protects plants from Pb toxicity by scavenging Pb-induced ROS generation (Liu et al. 2008). Pb has also been reported to induce several GSH genes such as glutathione synthetase, reductase, peroxidases, etc. In response to elevated levels of glutathione, the proline levels are also enhanced in stressed plants (Liu et al. 2008). Another report by Gupta et al. (2010) suggests the role of GSH in detoxification of Pb in *Sedum alfredii* plants without induction of PCs. This suggests that GSH play imperative role in Pb detoxification, where PCs are not present. Pb is known to induce the synthesis of PC and activates the enzyme PCS, i.e., phytochelatin synthetase (Andra et al. 2010; Singh et al. 2010) Moreover, Yadav et al. (2010) suggested involvement of in vivo PCs in cellular detoxification and sequestration of various heavy metals including Pb, possibly due to their ability to form Pb-PC complexes. PCs bind to soluble Pb ions in the cytoplasm and transfer them to chloroplast and vacuoles (Jiang and Liu 2010), thus resulting in lowering of toxic effects of Pb ions in plant cells.

5.4 *Metal-Chelating Compounds and Osmolytes*

In order to protect plants against Pb toxicity, plants have developed certain tolerance mechanisms including (1) pumping out of Pb at cell surface; (2) reduction in uptake of Pb; (3) chelation of Pb ions with thiol compounds in the cytosol, such as glutathione, phytochelatins, and metallothioneins; (4) sequestration in the inactive organelles such as vacuoles; and (5) detoxification of Pb-induced ROS generation (Kim et al. 2006; Verma and Dubey 2003; Zafari et al. 2016). Chelation is considered as one of the most significant mechanisms to counter heavy metal toxicity. The increase in contents of metal-chelating compounds occurs in response to activation of sulfur assimilation cascade under stressed conditions and eventually results in elevation in content of GSH (Mishra et al. 2009). In response to increased level of GSH, the synthesis of metal-chelating compounds specifically PCs is also enhanced (Ederli et al. 2004).

Heavy metal ion chelation with high-affinity ligands is a primary mechanism for metal detoxification. Phytochelatins (PCs) are characterized as biomarkers of early detection of metal toxicity in plants (Hossain et al. 2012). They are further categorized into nonprotein thiols and GSH groups and other –SH group-containing compounds (Kafel et al. 2010). The synthesis of PCs in the cytosol is triggered by the presence of several toxic metals Hg, Ag, Ni, Pb, Cu, As, and Au; however Cd is considered as the most important inducer (Grill et al. 1989; Vatamaniuk et al. 2001). The synthesized PCs are then transported in the form of M-PC (metal-phytochelatin) complexes (Manara 2012; Song et al. 2014). Increase in contents of PCs was recorded in response to Pb treatment (10^{-6} and 10^{-4} mol L^{-1}) in *Chlorella vulgaris*. Li et al. (2016) recorded a decline in levels of nonprotein sulfhydryls and phytochelatins in *Conyza canadensis* in response to Pb treatment. Elevation in levels of total acid thiols content was observed by 0.1, 1, 5, 10, and 20 μM Pb treatment in *Sesbania exaltata* L. plants (McComb et al. 2012). Nonprotein thiol content was significantly enhanced by Pb exposure in *Medicago sativa* (Pratima and Pratima 2016) and *Triticum aestivum* (Tripathi et al. 2016).

In order to regulate the osmotic balance, plants have specific protection strategy termed as osmoregulation which is mediated by osmolytes such as sugars, betaines, prolines, etc. These biomolecules aid in maintenance of turgor pressure of the cells under stress conditions (Misra and Saxena 2009). Osmoprotectants are of very low molecular weight and are organic in nature. They are nontoxic in nature and are essential for regulation of osmotic gradient of plants (Slama et al. 2015). Few reports suggest exogenous application of osmolytes to combat different stresses (Farooq et al. 2013). To maintain the driving gradient of the cytosol, the osmotic pressure was enhanced, eventually modulating water uptake and turgor pressure of cells (Serraj and Sinclair 2002). In addition to osmotic adjustment, these osmoprotectants have an ability to play an imperative role in ROS scavenging under stress conditions (Serraj and Sinclair 2002). Several sugars have been reported to elevate in response to abiotic stresses. They play a protective role in plants by modulating osmotic balance, membrane stability, and several other regulatory functions in plants under these conditions (Lokhande and Suprasanna 2012). Osmolytes also play a protective role in damage caused by dehydration and injury (Sharma and Dietz 2006; Ashraf and Foolad 2007). Various osmolytes include sugars (glucose, sucrose, fructose, trehalose, raffinose) and amino acids such as proline and alanine betaine, proline betaine, and polyamines (Mudgal et al. 2010). GB is an effective osmotic pressure under osmotic, metal, and salt stresses and is also an imperative compatible solute (Bharwana et al. 2014; Khan et al. 2015; Slama et al. 2015). Few reports of alteration in osmolyte contents in plants under Pb stress have been tabulated in Table 4.

5.5 *Role of PGRs*

Plant growth regulators (PGRs) form an important class of heavy metal-detoxifying agents (Clouse 2011). They have a wide range of applications in modern agriculture and management. PGRs have been adversely reported to enhance tolerance to various abiotic and biotic stresses (Anuradha et al. 2011). Several environmental conditions affect the content or endogenous levels of these phytohormones consequently modulating several signaling cascades. These changes induce negative

Table 4 Effect of Pb exposure on osmolyte content in different plant species

S. no.	Pb concentrations	Plant species	Effects on content of osmolytes	Reference
1.	10 and 100 μM	*Medicago sativa*	Significant enhancement in proline content	Pratima and Pratima (2016)
2.	100, 200, 400, and 800 ppm	*Arachis hypogea* (cultivar K6 and K9)	Enhancement in proline accumulation in both the cultivars	Nareshkumar et al. (2015)
3.	1.5, 3, and 15 mM	*Triticum aestivum* and *Spinacia oleracea*	Increase in level of proline was observed in leaves in both plants	Lamhamdi et al. (2013)
4.	100, 200, 400, and 800 ppm	*Brassica juncea arawali*	Total sugar content and proline content were enhanced	Kaur et al. (2013)
5.	100, 200, and 400 μM L^{-1}	*Brassica napus*	Increase in content of proline was recorded in both the varieties in roots and aerial parts also	Mosavian and Chaab (2012)
6.	0.05, 0.1, 0.5, and 1 g L^{-1}	*Triticum aestivum*	Proline content was significantly elevated	Lamhamdi et al. (2010)
7.	2.5 mM	*Raphanus sativus*	Contents of total sugars and reducing sugars were lowered	Anuradha et al. (2011)
8.	1, 10, 50, and 100 μM	*Najus indica*	Proline content was initially elevated followed by reduction in its content	Singh et al. (2010)
9.	0.5 mM	*Raphanus sativus*	Significant elevation in levels of proline was observed in leaves	Teklić et al. (2008)
10.	25, 50, 100, 200, 300, and 400 ppm	*Avena sativa*	Elevation in levels of total sugars and reducing sugars	Bhushan and Gupta (2008)

impact on plant growth and development under stressful environment conditions. Since the past decade, several studies suggest the role of exogenously applied plant hormones to cope with the stressful conditions (Ashraf 2009). Natural and synthetic phytohormones are exogenously applied either during growth of plant or prior to sowing, i.e., soaking treatment in phytohormones. Numerous studies have been documented suggesting imperative role of phytohormones in improving tolerance to metal stress (Choudhary et al. 2010, 2011; Kanwar and Bhardwaj 2015) (Fig. 4).

6 Techniques of Determination of Pb in Soil and Plants

Over the last few decades, quantification of metal ions in plants and other living system has been done using AAS, either graphite or flame. This technique enables the estimation of average total metal content in digested samples in parts per million (ppm), i.e., in ug L^{-1} (Carter et al. 2014). Pb quantification in soil and plants

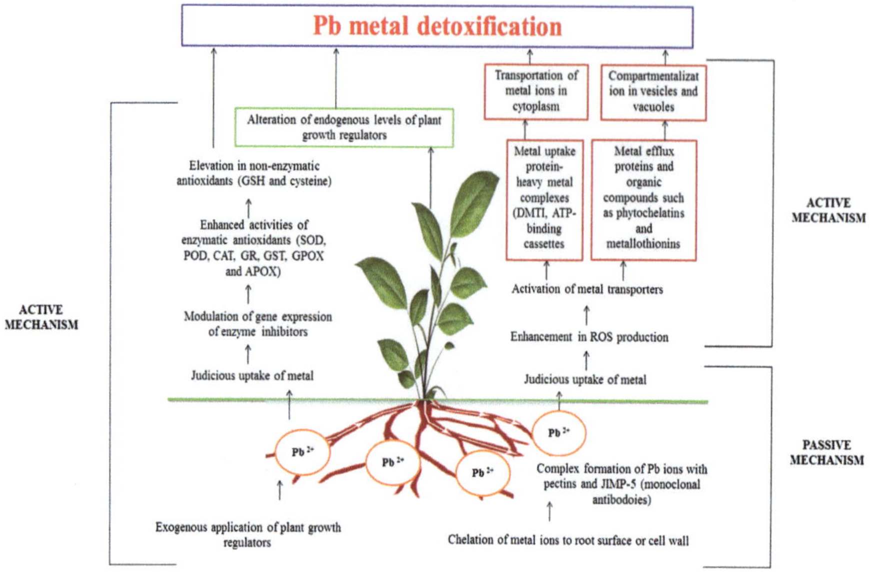

Fig. 4 Mechanism of Pb metal detoxification in plants

employing AAS has been largely documented (Rêgo et al. 2012; Zhong et al. 2016; Coco et al. 2017). Techniques such as ICP-MS and ICP-AES have comparatively higher sensitivity of 0.1–10 part per trillion (ppt), i.e., ng L^{-1} (Ammann 2007). ICP-MS and ICP-AES have been used for quantification of Pb in biological samples extensively (Ducos et al. 2010; Cho and Myung 2011; Zhao et al. 2018). Preparation of samples for quantification of Pb in soil as well as plant samples employing AAS, ICP-MS, and ICP-AES is done by initially drying the samples, followed by digesting in analytical grade nitric acid and perchloric acid in a ratio of 2:1. The digested samples are then diluted, filtered with Whatman No. 1 filter paper and further used for quantification of Pb (Allen et al. 1976).

Furthermore, elemental mapping of metals is a more recent approach for quantification of distribution of metals in biological samples. This approach utilizes instruments such as laser ablation coupled with ICP-MS (LA-ICP-MS), secondary ion mass spectrometry (SIMS), and nano-SIMS (Quinn et al. 2011; Moore et al. 2012). Fluorescent sensors have been designed to localize and visualize the accessible or labile pool of metal ions, i.e., free hydrated or loosely bound or buffered metal ions (Carter et al. 2014). A number of fluorescent sensors have been identified for localization of Pb in living system. These techniques consist of fluorophores and peptides, proteins, DNA duplexes, and nanoparticles. Leadfluor-1 is Pb-selective dicarboxylate pseudocrown ether and fluorescein-like xanthonone reporters which undergo an 18-fold fluorescence enhancement upon binding with Pb (He et al. 2006; Miller et al. 2008). Leadfluor-1 has also been used to localize Pb in plants (Miller et al. 2008; Karak et al. 2013). Leadmium green is another commercially available fluorescent sensor which becomes fluorescent even in the presence of nanomolar levels of Pb and micromolar levels of cadmium in plants (Jiang et al. 2014; Gao et al. 2015). Another report of use of a genetically encoded FRET (foster resonance energy transfer)-based Pb^{2+} sensor, i.e., MET lead 1.59, has been developed to detect Pb (Chereddy et al. 2015). Pb localization in plants employing FRET has been largely reported (Lewis 2010; Reddy et al. 2017; Terryn et al. 2018).

7 Conclusion and Future Prospects

This review includes an overview of the literature about Pb toxicity in plants. The insight into the basic mechanisms involved in lead uptake, translocation, sequestration, and toxicity caused in plants is extremely necessary for the understanding of detoxification and remediation of lead-contaminated environment. Exploration of modern molecular and genetic approaches to unravel biochemical and molecular alterations caused in response to metal stress in plants and identification of genes of metal chelators and transporters involved in detoxification of metal ions in plants system is imperative. These approaches might be useful in development of more tolerant transgenic plants for environmental cleanup and restoration of soil quality.

8 Summary

Pb is one of the most prominent examples of anthropogenic environmental pollutants. Agricultural soils contaminated with Pb result in drastic decline in crop productivity and have emerged as a serious problem toward agriculture. Pb is a nonessential element, and a large amount of it is easily absorbed by plants from the soil and is sequestered in the roots, and a very small amount is transported to the shoots. Primary sources of Pb include clouds from exhaust automobiles and chimneys of industries. Use of Pb in certain industries such as battery manufacturing, Pb ore extraction, and electroplating is also one of the reasons. Also mining and smelting as well as finishing operations have led to aggravation in Pb content in soils. In several developing countries, the use of Pb in paints, water pipes, cosmetic, acid batteries, and metal cans is common sources of addition of Pb to environment. Pb is available to plants from soils and aerosol sources. The range of soil Pb content is from 400 to 800 mg kg^{-1} of soil in Pb-contaminated soils. First visual symptom of Pb toxicity in plants includes speedy inhibition of root growth and development, retardation in growth, and roots turning black in color and showing symptoms of chlorosis. Inhibition of photosynthesis by Pb infiltration in plant cells further results in imbalance in nutrient uptake and water status as well as alteration in enzyme activities. Direct effect of Pb treatment has been observed on flowering plants where plants produce lesser number of flowers. Vital enzymes of chlorophyll biosynthesis pathway specifically α-amino levulinate dehydrogenase are strongly inhibited by Pb toxicity. Other photosynthetic and respiratory enzymes including ribulose-bisphosphate carboxylase/oxygenase, ATPases, etc. are also powerfully lowered in the presence of Pb toxicity.

HM triggers formation of ROS, caused by oxidative stress in plants. For ameliorating damaging effect of free radicals, plants have adopted certain strategies to regulate HM uptake, translocation, sequestration, and detoxification. The strategies are largely classified into detoxification and avoidance mechanisms. ROS can also be produced indirectly via (1) interaction of metal ions with antioxidant defense system, (2) replacement of essential metal ions with nonessential ions, and (3) disrupted electron transport chain leading to altered metabolism. ROS production results in degradation of lipid peroxidation and deterioration of the membranes. For scavenging the excessively produced ROS, plants produce certain antioxidative enzymes including SOD, CAT, POD, APOX, and GPOX and certain nonenzymatic antioxidants including ascorbic acid, glutathione, tocopherol, thioredoxins, etc. Antioxidant enzymes are a pool of glutathione that regulate the Asada-Halliwell pathways. In addition to this, plants also adopt certain avoidance strategies including (1) accumulation of few metal-chelating compounds like PCs, MTs, thiols (protein-bound and nonprotein-bound thiols), organic acids, and cysteine residues, (2) enhancement in content of phenolic compounds such as anthocyanins and flavonoids, and (3) aggravation in levels of osmolytes including sugars and amino acids. Several plant metabolites especially phytohormones are involved in protection of plants against oxidative stress produced by variable abiotic factors. These hormones include auxins, ABA, SA, GB, ethylene, and BRs.

Acknowledgment Financial help for carrying out above work was given by the University Grant Commission, Government of India, GOI (Maulana Azad National Fellowship), and DST-FIST, of GOI, is also duly acknowledged.

Contribution of Authors Sukhmeen Kaur Kohli, Renu Bhardwaj, and Saroj Arora designed the layout of the review article. Neha Handa, Shagun Bali, Kanika Khanna, and Anket Sharma helped in writing of the different sections of the manuscript. Renu Bhardwaj, Sukhmeen Kaur Kohli, and Kanika Khanna revised the manuscript to present form.

References

Agency for Toxic Substances and Disease Registry (2007) CERCLA priority list of hazardous substances. ATSDR Home. Retrieved March 22, 2011. From http://www.atsdr.cdc.gov/cercla/07list.html

Aggarwal A, Sharma I, Tripathi BN, Munjal AK, Baunthiyal M, Sharma V (2012) Metal toxicity and photosynthesis. In: Itoh S, Mohanty P, Guruprasad KN (eds.) Photosynthesis: Overviews on Recent Progress and Future Perspectives. IK International Publishing House (Pvt.) Limited, New Delhi, pp 229–236

Allen SE, Grimshaw HM, Parkinson JA, Quarmby C, Roberts JD (1976) Chemical analysis. In: Chapman SB (ed) Methods in plant ecology. Blackwell, Oxford, pp 424–426

Alves JD, Souza AP, Pôrto ML, Fontes RL, Arruda J, Marques LFH (2016) Potential of sunflower, castor bean, common buckwheat and vetiver as lead phytoaccumulators. Rev Bras Eng Agric Amb 20(3):243–249

Ammann AA (2007) Inductively coupled plasma mass spectrometry (ICP MS): a versatile tool. J Mass Spectrom 42(4):419–427

Andra SS, Datta R, Sarkar D, Makris KC, Mullens CP, Sahi SV, Bach SB (2010) Synthesis of phytochelatins in vetiver grass upon lead exposure in the presence of phosphorus. Plant Soil 326(1–2):171–185

Anuradha S, Rao R, Eeta S (2011) Amelioration of lead toxicity in radish (*Raphanus sativus* L) plants by brassinolide. J Appl Biol Sci 5(3):43–48

Arazi T, Sunkar R, Kaplan B, Fromm H (1999) A tobacco plasma membrane calmodulin-binding transporter confers Ni2+ tolerance and Pb2+ hypersensitivity in transgenic plants. Plant J 20(2):171–182

Arias JA, Peralta-Videa JR, Ellzey JT, Ren M, Viveros MN, Gardea-Torresdey JL (2010) Effects of *Glomus deserticola* inoculation on Prosopis: enhancing chromium and lead uptake and translocation as confirmed by X-ray mapping, ICP-OES and TEM techniques. Environ Exp Bot 68(2):139–148

Arshad M, Silvestre J, Pinelli E, Kallerhoff J, Kaemmerer M, Tarigo A, Shahid M, Guiresse M, Pradère P, Dumat C (2008) A field study of lead phytoextraction by various scented *Pelargonium* cultivars. Chemosphere 71(11):2187–2192

Arshad T, Maqbool N, Javed F, Arshad MU (2017) Enhancing the defensive mechanism of lead affected barley (*Hordeum vulgare* L.) genotypes by exogenously applied salicylic acid. J Agric Sci 9(2):139–146

Ashraf M (2009) Biotechnological approach of improving plant salt tolerance using antioxidants as markers. Biotechnol Adv 27(1):84–93

Ashraf M, Foolad M (2007) Roles of glycine betaine and proline in improving plant abiotic stress resistance. Environ Exp Bot 59(2):206–216

Ashraf M, Harris PJC (2013) Photosynthesis under stressful environments: an overview. Photosynthetica 51(2):163–190

Ashraf U, Kanu AS, Deng Q, Mo Z, Pan S, Tian H, Tang X (2017) Lead (Pb) toxicity; physio-biochemical mechanisms, grain yield, quality, and pb distribution proportions in scented rice. Front Plant Sci 8:259

Atici Ö, Ağar G, Battal P (2005) Changes in phytohormone contents in chickpea seeds germinating under lead or zinc stress. Biol Plant 49(2):215–222

Azhar NAZILA, Ashraf MY, Hussain M, Hussain F (2006) Phytoextraction of lead (Pb) by EDTA application through sunflower (*Helianthus annuus* L.) cultivation: seedling growth studies. Pak J Bot 38(5):1551–1560

Barceló J, Poschenrieder C (1990) Plant water relations as affected by heavy metal stress: a review. J Plant Nutr 13(1):1–37

Barrutia O, Garbisu C, Hernández-Allica J, García-Plazaola JI, Becerril JM (2010) Differences in EDTA-assisted metal phytoextraction between metallicolous and non-metallicolous accessions of *Rumex acetosa* L. Environ Pollut 158(5):1710–1715

Bharwana SA, Ali S, Farooq MA, Iqbal N, Abbas F, Ahmad MSA (2013) Alleviation of lead toxicity by silicon is related to elevated photosynthesis, antioxidant enzymes suppressed lead uptake and oxidative stress in cotton. J Bioremed Biodegr 4(4):4153–4172

Bharwana SA, Ali S, Farooq MA, Iqbal N, Hameed A, Abbas F, Ahmad MSA (2014) Glycine betaine-induced lead toxicity tolerance related to elevated photosynthesis, antioxidant enzymes suppressed lead uptake and oxidative stress in cotton. Turk J Bot 38(2):281–292

Bharwana SA, Shafaqat A, Farooq MA, Mujahid F, Nusrat B, Rehan A (2016) Physiological and biochemical changes induced by lead stress in cotton (*Gossypium hirsutum* L.) seedlings. Acad J Agric Res 4(4):160–167

Bhattacharjee S (2005) Reactive oxygen species and oxidative burst: roles in stress, senescence and signal transducation in plants. Curr Sci 89:1113–1121

Bhatti KH, Anwar S, Nawaz K, Hussain K (2013) Effect of heavy metal lead (Pb) stress of different concentration on wheat (*Triticum aestivum* L.). Middle-East J Sci Res 14(2):148–154

Bhushan B, Gupta K (2008) Effect of lead on carbohydrate mobilization in oat seeds during germination. J Appl Sci Environ Manag 12(2):29–33

Bi X, Ren L, Gong M, He Y, Wang L, Ma Z (2010) Transfer of cadmium and lead from soil to mangoes in an uncontaminated area, Hainan Island, China. Geoderma 155(1):115–120

Blaylock MJ (2000) Field demonstrations of phytoremediation of lead-contaminated soils. In: Phytoremediation of contaminated soil and water. Lewis, Boca Raton

Bressler JP, Olivi L, Cheong JH, Kim Y, Bannona D (2004) Divalent metal transporter 1 in lead and cadmium transport. Ann N Y Acad Sci 1012(1):142–152

Brunet J, Varrault G, Zuily-Fodil Y, Repellin A (2009) Accumulation of lead in the roots of grass pea (*Lathyrus sativus* L.) plants triggers systemic variation in gene expression in the shoots. Chemosphere 77(8):1113–1120

Carter KP, Young AM, Palmer AE (2014) Fluorescent sensors for measuring metal ions in living systems. Chem Rev 114(8):4564–4601

Chatterjee C, Dube BK, Sinha P, Srivastava P (2004) Detrimental effects of lead phytotoxicity on growth, yield, and metabolism of rice. Commun Soil Sci Plant Anal 35(1–2):255–265

Chen Q, Zhang X, Liu Y, Wei J, Shen W, Shen Z, Cui J (2017) Hemin-mediated alleviation of zinc, lead and chromium toxicity is associated with elevated photosynthesis, antioxidative capacity; suppressed metal uptake and oxidative stress in rice seedlings. Plant Growth Regul 81(2):253–264

Chereddy NR, Nagaraju P, Raju MN, Krishnaswamy VR, Korrapati PS, Bangal PR, Rao VJ (2015) A novel FRET 'off–on' fluorescent probe for the selective detection of Fe3+, Al3+ and Cr3+ ions: its ultrafast energy transfer kinetics and application in live cell imaging. Biosens Bioelectron 68:749–756

Cho HJ, Myung SW (2011) Determination of cadmium, chromium and lead in polymers by icp-oes using a high pressure asher (hpa). Bull Kor Chem Soc 32(2):489–497

Choudhary SP, Bhardwaj R, Gupta BD, Dutt P, Gupta RK, Kanwar M, Dutt P (2010) Changes induced by Cu2+ and Cr6+ metal stress in polyamines, auxins, abscisic acid titers and antioxidative enzymes activities of radish seedlings. Braz J Plant Physiol 22(4):263–270

Choudhary SP, Kanwar M, Bhardwaj R, Gupta BD, Gupta RK (2011) Epibrassinolide ameliorates Cr (VI) stress via influencing the levels of indole-3-acetic acid, abscisic acid, polyamines and antioxidant system of radish seedlings. Chemosphere 84(5):592–600

Clouse SD (2011) Brassinosteroid signal transduction: from receptor kinase activation to transcriptional networks regulating plant development. Plant Cell 23(4):1219–1230

Coco J, Bechlin MA, Barros AI, Ferreira EC, Veiga MA, Neto JA (2017) Development of analytical procedures for lead determination in incense by graphite furnace AAS. Atomic Spectrosc 1:208–212

Cotter-Howells J, Thornton I (1991) Sources and pathways of environmental lead to children in a Derbyshire mining village. Environ Geochem Health 13(2):127–135

Dai J, Mumper RJ (2010) Plant phenolics: extraction, analysis and their antioxidant and anticancer properties. Molecules 15(10):7313–7352

Decker A (1997) Phenolics: prooxidants or antioxidants? Nutr Rev 55(11):396–398

Devi PB, Vijayabharathi R, Sathyabama S, Malleshi NG, Priyadarisini VB (2014) Health benefits of finger millet (*Eleusine coracana* L.) polyphenols and dietary fiber: a review. J Food Sci Technol 51(6):1021–1040

Dey SK, Dey J, Patra S, Pothal D (2007) Changes in the antioxidative enzyme activities and lipid peroxidation in wheat seedlings exposed to cadmium and lead stress. Braz J Plant Physiol 19 (1):53–60

Dixon RA, Paiva NL (1995) Stress-induced phenylpropanoid metabolism. Plant Cell 7(7):1085

Dogan M, Karatas M, Aasim M (2018) Cadmium and lead bioaccumulation potentials of an aquatic macrophyte *Ceratophyllum demersum* L.: a laboratory study. Ecotoxicol Environ Saf 148:431–440

Ducos S, Hamester M, Godula M (2010) ICP-MS for detecting heavy metals in foodstuffs: the technology can analyze 50 samples in an hour. Food Quality Safety Magazine, Hoboken, USA

Dumat C, Quenea K, Bermond A, Toinen S, Benedetti MF (2006) Study of the trace metal ion influence on the turnover of soil organic matter in cultivated contaminated soils. Environ Pollut 142(3):521–529

Đurđević B, Lisjak M, Stošić M, Engler M, Popović B (2008) Influence of Pb and Cu toxicity on lettuce photosynthetic pigments and dry matter accumulation. Cereal Res Commun 36:1951–1954

Ederli L, Reale L, Ferranti F, Pasqualini S (2004) Responses induced by high concentration of cadmium in *Phragmites australis* roots. Physiol Plant 121(1):66–74

EPA (2001) Lead safe yards: developing and implementing a monitoring, assessment, and outreach program for your community. U.S. EPA Office of Research and Development, EPA/625/R-00/012

EPA (2017) United States Environmental Protection Agency: justification of appropriation estimates for the committee on appropriations. U.S. EPA Office of Research and Development, EPA/190/K-17/002

Fargasova A (2004) Toxicity comparison of some possible toxic metals (Cd, Cu, Pb, Se, Zn) on young seedlings of *Sinapis alba* L. Plant Soil Environ 50(1):33–38

Farooq MA, Ali S, Hameed A, Ishaque W, Mahmood K, Iqbal Z (2013) Alleviation of cadmium toxicity by silicon is related to elevated photosynthesis, antioxidant enzymes; suppressed cadmium uptake and oxidative stress in cotton. Ecotoxicol Environ Saf 96:242–249

Farooqi ZR, Iqbal MZ, Kabir M, Shafiq M (2009) Toxic effects of lead and cadmium on germination and seedling growth of *Albizia lebbeck* (L.) *Benth.* Pak J Bot 41(1):27–33

Flora SJS, Flora GJS, Saxena G (2006) Environmental occurrence, health effects and management of lead poisoning. In: Cascas SB, Sordo J (eds) Lead: chemistry, analytical aspects, environmental impacts and health effects. Elsevier, Amsterdam, pp 158–228

Gao W, Nan T, Tan G, Zhao H, Tan W, Meng F, Li Z, Li QX, Wang B (2015) Cellular and subcellular immunohistochemical localization and quantification of cadmium ions in wheat (*Triticum aestivum*). PLoS One 10(5):e0123779

Garg N, Aggarwal N (2011) Effects of interactions between cadmium and lead on growth, nitrogen fixation, phytochelatin, and glutathione production in mycorrhizal *Cajanus cajan* (L.) Mill sp. J Plant Growth Regul 30(3):286–300

Ghani A (2010) Toxic effects of heavy metals on plant growth and metal accumulation in maize (*Zea mays* L.). Iran J Toxicol 4(3):325–334

Ghani A, Khan I, Ahmed I, Mustafa I, Abd-Ur R, Muhammad N (2015) Amelioration of lead toxicity in *Pisum sativum* (L.) by foliar application of salicylic acid. J Environ Anal Toxicol 5:292

Gichner T, Žnidar I, Száková J (2008) Evaluation of DNA damage and mutagenicity induced by lead in tobacco plants. Mutat Res 652(2):186–190

Godbold DL, Kettner C (1991) Lead influences root growth and mineral nutrition of *Picea abies* seedlings. J Plant Physiol 139(1):95–99

Godzik A (1996) The structural alignment between two proteins: is there a unique answer? Protein Sci 5(7):1325–1338

Gopal R, Rizvi AH (2008) Excess lead alters growth, metabolism and translocation of certain nutrients in radish. Chemosphere 70(9):1539–1544

Grill E, Löffler S, Winnacker EL, Zenk MH (1989) Phytochelatins, the heavy-metal-binding peptides of plants, are synthesized from glutathione by a specific γ-glutamylcysteine dipeptidyl transpeptidase (phytochelatin synthase). Proc Natl Acad Sci 86(18):6838–6842

Grover P, Rekhadevi PV, Danadevi K, Vuyyuri SB, Mahboob M, Rahman MF (2010) Genotoxicity evaluation in workers occupationally exposed to lead. Int J Hyg Environ Health 213(2):99–106

Gupta DK, Srivastava A, Singh VP (2006) Phytoremediation of induced lead toxicity in *Vigna mungo* (L.) hepper by vetiver grass. Rohilkhand University, Bareilly

Gupta DK, Nicoloso FT, Schetinger MRC, Rossato LV, Pereira LB, Castro GY, Srivastava S, Tripathi RD (2009) Antioxidant defense mechanism in hydroponically grown *Zea mays* seedlings under moderate lead stress. J Hazard Mater 172(1):479–484

Gupta DK, Huang HG, Yang XE, Razafindrabe BHN, Inouhe M (2010) The detoxification of lead in *Sedum alfredii* H. is not related to phytochelatins but the glutathione. J Hazard Mater 177(1):437–444

Gurer H, Ercal N (2000) Can antioxidants be beneficial in the treatment of lead poisoning? Free Radic Biol Med 29(10):927–945

Harpaz-Saad S, Azoulay T, Arazi T, Ben-Yaakov E, Mett A, Shiboleth YM, Hörtensteiner S, Gidoni D, Gal-On A, Goldschmidt EE, Eyal Y (2007) Chlorophyllase is a rate-limiting enzyme in chlorophyll catabolism and is posttranslationally regulated. Plant Cell 19(3):1007–1022

Hassan M, Mansoor S (2014) Oxidative stress and antioxidant defense mechanism in mung bean seedlings after lead and cadmium treatments. Turk J Agric For 38(1):55–61

Hattab S, Hattab S, Flores-Casseres ML, Boussetta H, Doumas P, Hernandez LE, Banni M (2016) Characterisation of lead-induced stress molecular biomarkers in *Medicago sativa* plants. Environ Exp Bot 123:1–12

He Q, Miller EW, Wong AP, Chang CJ (2006) A selective fluorescent sensor for detecting lead in living cells. J Am Chem Soc 128(29):9316–9317

Hettiarachchi GM, Pierzynski GM (2004) Soil lead bioavailability and in situ remediation of lead-contaminated soils: a review. Environ Prog 23(1):78–93

Hossain Z, Hajika M, Komatsu S (2012) Comparative proteome analysis of high and low cadmium accumulating soybeans under cadmium stress. Amino Acids 43(6):2393–2416

Huang JW, Cunningham SD (1996) Lead phytoextraction: species variation in lead uptake and translocation. New Phytol 134(1):75–84

Hussain A, Abbas N, Arshad F, Akram M, Khan ZI, Ahmad K, Mansha M, Mirzaei F (2013) Effects of diverse doses of Lead (Pb) on different growth attributes of *Zea mays* L. Agric Sci 4(5):262

Islam E, Yang X, Li T, Liu D, Jin X, Meng F (2007) Effect of Pb toxicity on root morphology, physiology and ultrastructure in the two ecotypes of *Elsholtzia argyi*. J Hazard Mater 147(3):806–816

Islam E, Liu D, Li T, Yang X, Jin X, Mahmood Q, Tian S, Li J (2008) Effect of Pb toxicity on leaf growth, physiology and ultrastructure in the two ecotypes of *Elsholtzia argyi*. J Hazard Mater 154(1):914–926

Jarvis MD, Leung DWM (2002) Chelated lead transport in *Pinus radiata*: an ultrastructural study. Environ Exp Bot 48(1):21–32

Jayasri MA, Suthindhiran K (2017) Effect of zinc and lead on the physiological and biochemical properties of aquatic plant *Lemna minor*: its potential role in phytoremediation. Appl Water Sci 7(3):1247–1253

Jiang W, Liu D (2010) Pb-induced cellular defense system in the root meristematic cells of *Allium sativum* L. BMC Plant Biol 10(1):40

Jiang N, Luo X, Zeng J, Yang Z, Zheng L, Wang S (2010) Lead toxicity induced growth and antioxidant responses in *Luffa cylindrica* seedlings. Int J Agric Biol 12(2):205–210

Jiang Z, Qin R, Zhang H, Zou J, Shi Q, Wang J, Jiang W, Liu D (2014) Determination of Pb genotoxic effects in Allium cepa root cells by fluorescent probe, microtubular immunofluorescence and comet assay. Plant Soil 383(1–2):357–372

Jing CHEN, Cheng ZHU, Li LP, Sun ZY, Pan XB (2007) Effects of exogenous salicylic acid on growth and H_2O_2-metabolizing enzymes in rice seedlings under lead stress. J Environ Sci 19(1):44–49

John R, Ahmad P, Gadgil K, Sharma S (2012) Heavy metal toxicity: effect on plant growth, biochemical parameters and metal accumulation by Brassica juncea L. Int J Plant Prod 3(3):65–76

Kabata-Pendias A (1993) Behavioural properties of trace metals in soils. Appl Geochem 8:3–9

Kafel A, Nadgórska-Socha A, Gospodarek J, Babczyńska A, Skowronek M, Kandziora M, Rozpędek K (2010) The effects of Aphis fabae infestation on the antioxidant response and heavy metal content in field grown *Philadelphus coronarius* plants. Sci Total Environ 408(5):1111–1119

Kanwar MK, Bhardwaj R (2015) Arsenic induced modulation of antioxidative defense system and brassinosteroids in *Brassica juncea* L. Ecotoxicol Environ Saf 115:119–125

Karak D, Banerjee A, Lohar S, Sahana A, Mukhopadhyay SK, Adhikari SS, Das D (2013) Xanthone based Pb^{2+} selective turn on fluorescent probe for living cell staining. Anal Methods 5(1):169–172

Kaur L, Gadgil K, Sharma S (2013) Comparative study of natural phyto-extraction and induced phyto-extraction of lead using mustard plant (*Brassica juncea arawali*). Int J Bioassays 2(1):352–357

Khan AL, Waqas M, Hussain J, Al-Harrasi A, Hamayun M, Lee IJ (2015) Phytohormones enabled endophytic fungal symbiosis improve aluminum phytoextraction in tolerant *Solanum lycopersicum*: an examples of *Penicillium janthinellum* LK5 and comparison with exogenous GA 3. J Hazard Mater 295:70–78

Khan M, Daud MK, Basharat A, Khan MJ, Azizullah A, Muhammad N, Muhammad N, ur Rehman Z, Zhu SJ (2016) Alleviation of lead-induced physiological, metabolic, and ultramorphological changes in leaves of upland cotton through glutathione. Environ Sci Pollut Res 23(9):8431–8440

Kim DY, Bovet L, Kushnir S, Noh EW, Martinoia E, Lee Y (2006) AtATM3 is involved in heavy metal resistance in *Arabidopsis*. Plant Physiol 140(3):922–932

Kohli SK, Handa N, Sharma A, Kumar V, Kaur P, Bhardwaj R (2017) Synergistic effect of 24-epibrassinolide and salicylic acid on photosynthetic efficiency and gene expression in *Brassica juncea* L. under Pb stress. Turk J Biol 41(6):943–953

Kohli SK, Handa N, Bali S, Arora S, Sharma A, Kaur R, Bhardwaj R (2018a) Modulation of antioxidative defense expression and osmolyte content by co-application of 24-epibrassinolide and salicylic acid in Pb exposed Indian mustard plants. Ecotoxicol Environ Saf 147:382–393

Kohli SK, Handa N, Sharma A, Gautam V, Arora S, Bhardwaj R, Wijaya L, Alyemeni MN, Ahmad P (2018b) Interaction of 24-epibrassinolide and salicylic acid regulates pigment contents, antioxidative defense responses, and gene expression in *Brassica juncea* L. seedlings under Pb stress. Environ Sci Pollut Res 25(15):15159–15173

Kopittke PM, Asher CJ, Kopittke RA, Menzies NW (2007) Toxic effects of Pb 2+ on growth of cowpea (*Vigna unguiculata*). Environ Pollut 150(2):280–287

Kos B, Greman H, Lestan D (2003) Phytoextraction of lead, zinc and cadmium from soil by selected plants. Plant Soil Environ 49(12):548–553

Krzesłowska M (2011) The cell wall in plant cell response to trace metals: polysaccharide remodeling and its role in defense strategy. Acta Physiol Plant 33(1):35–51

Krzesłowska M, Lenartowska M, Mellerowicz EJ, Samardakiewicz S, Woźny A (2009) Pectinous cell wall thickenings formation—a response of moss protonemata cells to lead. Environ Exp Bot 65(1):119–131

Krzesłowska M, Lenartowska M, Samardakiewicz S, Bilski H, Woźny A (2010) Lead deposited in the cell wall of *Funaria hygrometrica* protonemata is not stable–a remobilization can occur. Environ Pollut 158(1):325–338

Kumar M, Jayaraman P (2014) Toxic effect of lead nitrate [$Pb(NO_3)_2$] on the black gram seedlings (*Vigna mungo* L. Hepper). Int J Adv Res Biol Sci 1(9):209–213

Kumar A, Pal L, Agrawal V (2017) Glutathione and citric acid modulates lead-and arsenic-induced phytotoxicity and genotoxicity responses in two cultivars of *Solanum lycopersicum* L. Acta Physiol Plant 39(7):151

Lamhamdi M, Bakrim A, Aarab A, Lafont R, Sayah F (2010) A comparison of lead toxicity using physiological and enzymatic parameters on spinach (*Spinacia oleracea*) and wheat (*Triticum aestivum*) growth. Moroccan J Biol 6(7):64–73

Lamhamdi M, El Galiou O, Bakrim A, Nóvoa-Muñoz JC, Arias-Estévez M, Aarab A, Lafont R (2013) Effect of lead stress on mineral content and growth of wheat (*Triticum aestivum*) and spinach (*Spinacia oleracea*) seedlings. Saudi J Biol Sci 20(1):29–36

Lane SD, Martin ES (1977) A histochemical investigation of lead uptake in *Raphanus sativus*. New Phytol 79(2):281–286

Lavid N, Schwartz A, Yarden O, Tel-Or E (2001) The involvement of polyphenols and peroxidase activities in heavy-metal accumulation by epidermal glands of the waterlily (Nymphaeaceae). Planta 212(3):323–331
Lawal OS, Sanni AR, Ajayi IA, Rabiu OO (2010) Equilibrium, thermodynamic and kinetic studies for the biosorption of aqueous lead (II) ions onto the seed husk of *Calophyllum inophyllum*. J Hazard Mater 177(1):829–835
Leal-Alvarado DA, Espadas-Gil F, Sáenz-Carbonell L, Talavera-May C, Santamaría JM (2016) Lead accumulation reduces photosynthesis in the lead hyper-accumulator Salvinia minima Baker by affecting the cell membrane and inducing stomatal closure. Aquat Toxicol 171:37–47
Lewis A (2010) Determination of lead levels in soil and plant uptake studies. Rev J Undergrad Stud Res 12(1):48–56
Li HY, Wei DQ, Shen M, Zhou ZP (2012) Endophytes and their role in phytoremediation. Fungal Divers 54(1):11–18
Li Y, Zhou C, Huang M, Luo J, Hou X, Wu P, Ma X (2016) Lead tolerance mechanism in *Conyza canadensis*: subcellular distribution, ultrastructure, antioxidative defense system, and phytochelatins. J Plant Res 129(2):251–262
Liao YC, Chien SC, Wang MC, Shen Y, Hung PL, Das B (2006) Effect of transpiration on Pb uptake by lettuce and on water soluble low molecular weight organic acids in rhizosphere. Chemosphere 65(2):343–351
Lindsay WL (1979) Chemical equilibria in soils. Wiley, New York
Liu D, Li TQ, Jin XF, Yang XE, Islam E, Mahmood Q (2008) Lead induced changes in the growth and antioxidant metabolism of the lead accumulating and non-accumulating ecotypes of *Sedum alfredii*. J Integr Plant Biol 50(2):129–140
Liu X, Peng K, Wang A, Lian C, Shen Z (2010) Cadmium accumulation and distribution in populations of *Phytolacca americana* L. and the role of transpiration. Chemosphere 78(9):1136–1141
Lokhande VH, Suprasanna P (2012) Prospects of halophytes in understanding and managing abiotic stress tolerance. In: Environmental adaptations and stress tolerance of plants in the era of climate change. Springer, New York, pp 29–56
Maestri E, Marmiroli M, Visioli G, Marmiroli N (2010) Metal tolerance and hyperaccumulation: costs and trade-offs between traits and environment. Environ Exp Bot 68(1):1–13
Malar S, Vikram SS, Favas PJ, Perumal V (2014) Lead heavy metal toxicity induced changes on growth and antioxidative enzymes level in water hyacinths [*Eichhornia crassipes* (Mart.)]. Bot Stud 55(1):54
Małecka A, Piechalak A, Morkunas I, Tomaszewska B (2008) Accumulation of lead in root cells of Pisum sativum. Acta Physiol Plant 30(5):629–637
Malone C, Koeppe DE, Miller RJ (1974) Localization of lead accumulated by corn plants. Plant Physiol 53(3):388–394
Manara A (2012) Plant responses to heavy metal toxicity. In: Plants and heavy metals. Springer, Dordrecht, pp 27–53
Manginsay-Enot M, Sillero-Mahinay M (2017) A study of potential lead metal hyperaccumulator in tissue cultured plants of Tradescantia spathacea Sw. and Chlorophytum orchidastrum grown in hydroponics. Bull Environ Pharmacol Life Sci 6:45–50
Maodzeka A, Hussain N, Wei L, Zvobgo G, Mapodzeke JM, Adil MF, Jabeen S, Wang F, Jiang L, Shamsi IH (2017) Elucidating the physiological and biochemical responses of different tobacco (Nicotiana tabacum) genotypes to lead toxicity. Environ Toxicol Chem 36(1):175–181
McComb J, Hentz S, Miller GS, Begonia M (2012) Effects of lead on plant growth, lead accumulation and phytochelatin contents of hydroponically-grown *Sesbania exaltata*. World Environ 2(3):38–43
Meyers DE, Auchterlonie GJ, Webb RI, Wood B (2008) Uptake and localisation of lead in the root system of *Brassica juncea*. Environ Pollut 153(2):323–332
Michalak A (2006) Phenolic compounds and their antioxidant activity in plants growing under heavy metal stress. Pol J Environ Stud 15(4):523–530

Miller G, Shulaev V, Mittler R (2008) Reactive oxygen signaling and abiotic stress. Physiol Plant 133(3):481–489

Mishra S, Srivastava S, Tripathi RD, Kumar R, Seth CS, Gupta DK (2006) Lead detoxification by coontail (*Ceratophyllum demersum* L.) involves induction of phytochelatins and antioxidant system in response to its accumulation. Chemosphere 65(6):1027–1039

Mishra S, Tripathi RD, Srivastava S, Dwivedi S, Trivedi PK, Dhankher OP, Khare A (2009) Thiol metabolism play significant role during cadmium detoxification by *Ceratophyllum demersum* L. Bioresour Technol 100(7):2155–2161

Misra N, Saxena P (2009) Effect of salicylic acid on proline metabolism in lentil grown under salinity stress. Plant Sci 177(3):181–189

Mohan BS, Hosetti BB (1997) Potential phytotoxicity of lead and cadmium to *Lemna minor* grown in sewage stabilization ponds. Environ Pollut 98(2):233–238

Mohtadi A, Ghaderian SM, Schat H (2012) A comparison of lead accumulation and tolerance among heavy metal hyperaccumulating and non-hyperaccumulating metallophytes. Plant Soil 352(1–2):267–276

Moore KL, Lombi E, Zhao FJ, Grovenor CR (2012) Elemental imaging at the nanoscale: NanoSIMS and complementary techniques for element localisation in plants. Anal Bioanal Chem 402(10):3263–3273

Mosavian SN, Chaab A (2012) Response of morphologic and physiologic two variety canola to stress concentration of lead. Adv Environ Biol 6:1503–1509

Mudgal V, Madaan N, Mudgal A, Singh R, Mishra S (2010) Effect of toxic metals on human health. Open Nutraceuticals J 3:94–99

Munzuroglu O, Geckil H (2002) Effects of metals on seed germination, root elongation, and coleoptile and hypocotyl growth in *Triticum aestivum* and *Cucumis sativus*. Arch Environ Contam Toxicol 43(2):203–213

Muszyńska E, Kałużny K, Hanus-Fajerska E (2014) Phenolic compounds in *Hippophaë rhamnoides* leaves collected from heavy metals contaminated sites.[W:] Plants in urban areas and landscape. Slovak University of Agriculture in Nitra, Faculty of Horticulture and Landscape Engineering, Nitra, pp 11–14

Nagajyoti PC, Lee KD, Sreekanth TVM (2010) Heavy metals, occurrence and toxicity for plants: a review. Environ Chem Lett 8(3):199–216

Nareshkumar A, Veeranagamallaiah G, Pandurangaiah M, Kiranmai K, Amaranathareddy V, Lokesh U, Venkatesh B, Sudhakar C (2015) Pb-stress induced oxidative stress caused alterations in antioxidant efficacy in two groundnut (*Arachis hypogaea* L.) cultivars. Agric Sci 6 (10):1283

Ojwang LO, Banerjee N, Noratto GD, Angel-Morales G, Hachibamba T, Awika JM, Mertens-Talcott SU (2015) Polyphenolic extracts from cowpea (*Vigna unguiculata*) protect colonic myofibroblasts (CCD18Co cells) from lipopolysaccharide (LPS)-induced inflammation–modulation of microRNA 126. Food Funct 6(1):145–153

Olivares E (2003) The effect of lead on the phytochemistry of Tithonia diversifolia exposed to roadside automotive pollution or grown in pots of Pb-supplemented soil. Braz J Plant Physiol 15(3):149–158

Padmavathiamma PK, Li LY (2010) Phytoavailability and fractionation of lead and manganese in a contaminated soil after application of three amendments. Bioresour Technol 101(14):5667–5676

Pais I, Jones JB Jr (1997) The handbook of trace elements. CRC Press, Boca Raton

Pan J, Lin S, Woodbury NW (2012) Bacteriochlorophyll excited-state quenching pathways in bacterial reaction centers with the primary donor oxidized. J Phys Chem B 116(6):2014–2022

Parys E, Romanowska E, Siedlecka M, Poskuta JW (1998) The effect of lead on photosynthesis and respiration in detached leaves and in mesophyll protoplasts of *Pisum sativum*. Acta Physiol Plant 20(3):313–322

Patra M, Bhowmik N, Bandopadhyay B, Sharma A (2004) Comparison of mercury, lead and arsenic with respect to genotoxic effects on plant systems and the development of genetic tolerance. Environ Exp Bot 52(3):199–223

Piechalak A, Tomaszewska B, Baralkiewicz D, Malecka A (2002) Accumulation and detoxification of lead ions in legumes. Phytochemistry 60(2):153–162

Pinho S, Ladeiro B (2012) Phytotoxicity by lead as heavy metal focus on oxidative stress. J Bot 20:1–10

Piotrowska A, Bajguz A, Godlewska-Żyłkiewicz B, Czerpak R, Kamińska M (2009) Jasmonic acid as modulator of lead toxicity in aquatic plant *Wolffia arrhiza* (Lemnaceae). Environ Exp Bot 66(3):507–513

Pourrut B, Perchet G, Silvestre J, Cecchi M, Guiresse M, Pinelli E (2008) Potential role of NADPH-oxidase in early steps of lead-induced oxidative burst in *Vicia faba* roots. J Plant Physiol 165(6):571–579

Pourrut B, Shahid M, Dumat C, Winterton P, Pinelli E (2011) Lead uptake, toxicity, and detoxification in plants. In: Reviews of environmental contamination and toxicology. Springer, New York, pp 113–136

Pratima M, Pratima H (2016) Copper toxicity causes oxidative stress in *Brassica juncea* L. seedlings. Indian J Plant Physiol 14(4):397–401

Punamiya P, Datta R, Sarkar D, Barber S, Patel M, Das P (2010) Symbiotic role of *Glomus mosseae* in phytoextraction of lead in vetiver grass [*Chrysopogon zizanioides* (L.)]. J Hazard Mater 177(1):465–474

Qian JIN, Shan XQ, Wang ZJ, Tu Q (2016) Distribution and plant availability of heavy metals in different particle-size fractions of soil. Sci Total Environ 187(2):131–141

Qufei L, Fashui H (2009) Effects of Pb2+ on the structure and function of photosystem II of *Spirodela polyrrhiza*. Biol Trace Elem Res 129(1–3):251

Quinn CF, Prins CN, Freeman JL, Gross AM, Hantzis LJ, Reynolds RJ, Covey PA, Bañuelos GS, Pickering IJ, Fakra SC, Marcus MA (2011) Selenium accumulation in flowers and its effects on pollination. New Phytol 192(3):727–737

Qureshi MI, Abdin MZ, Qadir S, Iqbal M (2007) Lead-induced oxidative stress and metabolic alterations in *Cassia angustifolia* Vahl. Biol Plant 51(1):121–128

Rao SSR, Raghu K (2016) Amelioration of lead toxicity by 24-epibrassinolide in rose-scented geranium [*Pelargonium graveolens* (L.) Herit], a potential agent for phytoremediation. J Med Plants 4(6):06–08

Reddy KE, Park KR, Lee SD, Yoo JH, Son AR, Lee HJ (2017) Effects of graded concentrations of supplemental lead on lead concentrations in tissues of pigs and prediction equations for estimating dietary lead intake. PeerJ 5:e3936

Rêgo JF, Virgilio A, Nobrega JA, Neto JA (2012) Determination of lead in medicinal plants by high-resolution continuum source graphite furnace atomic absorption spectrometry using direct solid sampling. Talanta 100:21–26

Rice-Evans C, Miller N, Paganga G (1997) Antioxidant properties of phenolic compounds. Trends Plant Sci 2(4):152–159

Romanowska E, Wróblewska B, Drozak A, Siedlecka M (2006) High light intensity protects photosynthetic apparatus of pea plants against exposure to lead. Plant Physiol Biochem 44(5):387–394

Rossato LV, Nicoloso FT, Farias JG, Cargnelluti D, Tabaldi LA, Antes FG, Dressler VL, Morsch VM, Schetinger MRC (2012) Effects of lead on the growth, lead accumulation and physiological responses of *Pluchea sagittalis*. Ecotoxicology 21(1):111–123

Sakihama Y, Yamasaki H (2002) Lipid peroxidation induced by phenolics in conjunction with aluminum ions. Biol Plant 45(2):249–254

Saleem M, Asghar HN, Zahir ZA, Shahid M (2018) Impact of lead tolerant plant growth promoting rhizobacteria on growth, physiology, antioxidant activities, yield and lead content in sunflower in lead contaminated soil. Chemosphere 195:606–614

Sandalio LM, Dalurzo HC, Gomez M, Romero-Puertas MC, Del Rio LA (2001) Cadmium-induced changes in the growth and oxidative metabolism of pea plants. J Exp Bot 52(364):2115–2126

Scandalios JG (2005) Oxidative stress: molecular perception and transduction of signals triggering antioxidant gene defenses. Braz J Med Biol Res 38(7):995–1014

Šebestík J, Marques SM, Falé PL, Santos S, Arduíno DM, Cardoso SM, Oliveira CR, Serralheiro MLM, Santos MA (2011) Bifunctional phenolic-choline conjugates as anti-oxidants and acetylcholinesterase inhibitors. J Enzyme Inhib Med Chem 26(4):485–497
Sengar AS, Thind KS, Kumar B, Pallavi M, Gosal SS (2009) In vitro selection at cellular level for red rot resistance in sugarcane (Saccharum sp.). Plant Growth Regul 58(2):201–209
Seregin IV, Ivanov VB (2001) Physiological aspects of cadmium and lead toxic effects on higher plants. Russ J Plant Physiol 48(4):523–544
Seregin IV, Shpigun LK, Ivanov VB (2004) Distribution and toxic effects of cadmium and lead on maize roots. Russ J Plant Physiol 51(4):525–533
Serraj R, Sinclair TR (2002) Osmolyte accumulation: can it really help increase crop yield under drought conditions? Plant Cell Environ 25(2):333–341
Shahid MA, Pervez MA, Balal RM, Mattson NS, Rashid A, Ahmad R, Ayyub CM, Abbas T (2011) Brassinosteroid (24-epibrassinolide) enhances growth and alleviates the deleterious effects induced by salt stress in pea (*'Pisum sativum'* L.). Aust J Crop Sci 5(5):500
Shahid M, Dumat C, Pourrut B, Abbas G, Shahid N, Pinelli E (2015) Role of metal speciation in lead-induced oxidative stress to *Vicia faba* roots. Russ J Plant Physiol 62(4):448–454
Sharma SS, Dietz KJ (2006) The significance of amino acids and amino acid-derived molecules in plant responses and adaptation to heavy metal stress. J Exp Bot 57(4):711–726
Sharma P, Dubey RS (2005) Lead toxicity in plants. Braz J Plant Physiol 17(1):35–52
Sharma I, Pati PK, Bhardwaj R (2011) Effect of 28-homobrassinolide on antioxidant defence system in *Raphanus sativus* L. under chromium toxicity. Ecotoxicology 20(4):862–874
Shu X, Yin L, Zhang Q, Wang W (2012) Effect of Pb toxicity on leaf growth, antioxidant enzyme activities, and photosynthesis in cuttings and seedlings of *Jatropha curcas* L. Environ Sci Pollut Res 19(3):893–902
Sidhu GPS, Singh HP, Batish DR, Kohli RK (2016) Effect of lead on oxidative status, antioxidative response and metal accumulation in *Coronopus didymus*. Plant Physiol Biochem 105:290–296
Sidhu GPS, Bali AS, Bhardwaj R, Singh HP, Batish DR, Kohli RK (2018) Bioaccumulation and physiological responses to lead (Pb) in Chenopodium murale L. Ecotoxicol Environ Saf 151:83–90
Silva S, Pinto G, Santos C (2017) Low doses of Pb affected Lactuca sativa photosynthetic performance. Photosynthetica 55(1):50–57
Singh R, Tripathi RD, Dwivedi S, Kumar A, Trivedi PK, Chakrabarty D (2010) Lead bioaccumulation potential of an aquatic macrophyte Najas indica are related to antioxidant system. Bioresour Technol 101(9):3025–3032
Sinha P, Dube BK, Srivastava P, Chatterjee C (2006) Alteration in uptake and translocation of essential nutrients in cabbage by excess lead. Chemosphere 65(4):651–656
Slama I, Abdelly C, Bouchereau A, Flowers T, Savouré A (2015) Diversity, distribution and roles of osmoprotective compounds accumulated in halophytes under abiotic stress. Ann Bot 115 (3):433–447
Song WY, Yamaki T, Yamaji N, Ko D, Jung KH, Fujii-Kashino M, An G, Martinoia E, Lee Y, Ma JF (2014) A rice ABC transporter, OsABCC1, reduces arsenic accumulation in the grain. Proc Natl Acad Sci 111(44):15699–15704
Srivastava D, Singh A, Baunthiyal M (2015) Lead toxicity and tolerance in plants. J Plant Sci Res 2(2):123
Stefanov K, Seizova K, Yanishlieva N, Marinova E, Popov S (1995) Accumulation of lead, zinc and cadmium in plant seeds growing in metalliferous habitats in Bulgaria. Food Chem 54(3):311–313
Tabelin CB, Igarashi TO (2009) Mechanisms of arsenic and lead release from hydrothermally altered rock. J Hazard Mater 169(1):980–990
Taiz L, Zeiger E (2010) Photosynthesis: the light reactions. Plant Physiol 5:163–198
Tang C, Song J, Hu X, Hu X, Zhao Y, Li B, Ou D, Peng L (2017) Exogenous spermidine enhanced Pb tolerance in *Salix matsudana* by promoting Pb accumulation in roots and spermidine, nitric oxide, and antioxidant system levels in leaves. Ecol Eng 107:41–48

Tangahu BV, Sheikh Abdullah SR, Basri H, Idris M, Anuar N, Mukhlisin M (2011) A review on heavy metals (As, Pb, and Hg) uptake by plants through phytoremediation. Int J Chem Eng 2011:31

Teklić T, Hancock JT, Engler M, Paradicović N, Cesar V, Lepeduš H, Štolfa I, Bešlo D (2008) Antioxidative responses in radish (*Raphanus sativus* L.) plants stressed by copper and lead in nutrient solution and soil. Acta Biol Cracov 50(2):79–86

Terryn C, Paës G, Spriet C (2018) FRET-SLiM on native autofluorescence: a fast and reliable method to study interactions between fluorescent probes and lignin in plant cell wall. Plant Methods 14(1):74

Tripathi DK, Singh VP, Prasad SM, Dubey NK, Chauhan DK, Rai AK (2016) LIB spectroscopic and biochemical analysis to characterize lead toxicity alleviative nature of silicon in wheat (*Triticum aestivum* L.) seedlings. J Photochem Photobiol B Biol 154:89–98

Uzu G, Sobanska S, Aliouane Y, Pradere P, Dumat C (2009) Study of lead phytoavailability for atmospheric industrial micronic and sub-micronic particles in relation with lead speciation. Environ Pollut 157(4):1178–1185

Vatamaniuk OK, Bucher EA, Ward JT, Rea PA (2001) A new pathway for heavy metal detoxification in animals phytochelatin synthase is required for cadmium tolerance in *Caenorhabditis elegans*. J Biol Chem 276(24):20817–20820

Vega FA, Andrade ML, Covelo EF (2010) Influence of soil properties on the sorption and retention of cadmium, copper and lead, separately and together, by 20 soil horizons: comparison of linear regression and tree regression analyses. J Hazard Mater 174(1):522–533

Verbruggen N, Hermans C, Schat H (2009) Molecular mechanisms of metal hyperaccumulation in plants. New Phytol 181(4):759–776

Verma S, Dubey RS (2003) Lead toxicity induces lipid peroxidation and alters the activities of antioxidant enzymes in growing rice plants. Plant Sci 164(4):645–655

Wang HH, Shan XQ, Wen B, Owens G, Fang J, Zhang SZ (2007) Effect of indole-3-acetic acid on lead accumulation in maize (*Zea mays* L.) seedlings and the relevant antioxidant response. Environ Exp Bot 61(3):246–253

WHO (2001) Regional Office for Europe, air quality guidelines, chapter 6.7, Lead, Copenhagen, Denmark, 2nd edn. http://www.euro.who.int/document/aiq/67lead.pdf

Wierzbicka M (1999) Comparison of lead tolerance in *Allium cepa* with other plant species. Environ Pollut 104(1):41–52

Wierzbicka MH, Przedpełska E, Ruzik R, Ouerdane L, Połeć-Pawlak K, Jarosz M, Szpunar J, Szakiel A (2007) Comparison of the toxicity and distribution of cadmium and lead in plant cells. Protoplasma 231(1):99–111

Williams JC (1976) The segregation of particulate materials. A review. Powder Technol 15(2):245–251

Winkel-Shirley B (2002) Biosynthesis of flavonoids and effects of stress. Curr Opin Plant Biol 5(3):218–223

Wiszniewska A, Muszyńska E, Hanus-Fajerska E, Smoleń S, Dziurka M, Dziurka K (2017) Organic amendments enhance Pb tolerance and accumulation during micropropagation of *Daphne jasminea*. Environ Sci Pollut Res 24(3):2421–2432

Wojas S, Ruszczyńska A, Bulska E, Wojciechowski M, Antosiewicz DM (2007) Ca^{2+}-dependent plant response to Pb^{2+} is regulated by LCT1. Environ Pollut 147(3):584–592

Wu SC, Luo YM, Cheung KC, Wong MH (2006) Influence of bacteria on Pb and Zn speciation, mobility and bioavailability in soil: a laboratory study. Environ Pollut 144(3):765–773

Xiong ZT (1997) Bioaccumulation and physiological effects of excess lead in a roadside pioneer species *Sonchus oleraceus* L. Environ Pollut 97(3):275–279

Xiong ZT, Wang H (2005) Copper toxicity and bioaccumulation in Chinese cabbage (*Brassica pekinensis* Rupr.). Environ Toxicol 20(2):188–194

Xu B, Wang Y, Zhang S, Guo Q, Jin Y, Chen J, Gao Y, Ma H (2017) Transcriptomic and physiological analyses of *Medicago sativa* L. roots in response to lead stress. PLoS One 12(4):e0175307

Yadav DV, Jain R, Rai RK (2010) Impact of heavy metals on sugar cane. In: Sherameti I, Varma A (eds) Soil heavy metals, Soil biology series. Springer, Berlin, pp 339–367

Yan ZZ, Ke L, Tam NFY (2010) Lead stress in seedlings of *Avicennia marina*, a common mangrove species in South China, with and without cotyledons. Aquat Bot 92(2):112–118

Yang YY, Jung JY, Song WY, Suh HS, Lee Y (2000) Identification of rice varieties with high tolerance or sensitivity to lead and characterization of the mechanism of tolerance. Plant Physiol 124(3):1019–1026

Zafari S, Sharifi M, Chashmi NA, Mur LA (2016) Modulation of Pb-induced stress in Prosopis shoots through an interconnected network of signaling molecules, phenolic compounds and amino acids. Plant Physiol Biochem 99:11–20

Zaier H, Ghnaya T, Ghabriche R, Chmingui W, Lakhdar A, Lutts S, Abdelly C (2014) EDTA-enhanced phytoremediation of lead-contaminated soil by the halophyte Sesuvium portulacastrum. Environ Sci Pollut Res 21(12):7607–7615

Zhao HJ, Ge ML, Yan Y, Zhang TJ, Zeng J, Zhou WY, Wang YF, Meng QH, Zhang CB (2018) Inductively coupled plasma mass spectrometry as a reference method to evaluate serum calcium measurement bias and the commutability of processed materials during routine measurements. Chin Med J 131(13):1584

Zhong WS, Ren T, Zhao LJ (2016) Determination of Pb (Lead), Cd (Cadmium), Cr (Chromium), Cu (Copper), and Ni (Nickel) in Chinese tea with high-resolution continuum source graphite furnace atomic absorption spectrometry. J Food Drug Anal 24(1):46–55

Index

P. de Voogt (ed.), *Reviews of Environmental Contamination and Toxicology, Volume 249*, Reviews of Environmental Contamination and Toxicology Volume 249,
https://doi.org/10.1007/978-3-030-20194-4

MIX
Papier aus verantwortungsvollen Quellen
Paper from responsible sources
FSC® C105338
FSC
www.fsc.org

If you have any concerns about our products,
you can contact us on
ProductSafety@springernature.com

In case Publisher is established outside the EU,
the EU authorized representative is:
Springer Nature Customer Service Center GmbH
Europaplatz 3, 69115 Heidelberg, Germany

Printed by Libri Plureos GmbH
in Hamburg, Germany